EVOLUTION OF MAN

Channeled by
THE SPIRITUAL HIERARCHY
through

NADA-YOLANDA

Published for
THE HIERARCHAL BOARD
and the
UNIVERSITY OF LIFE
by
MARK-AGE
Ft. Lauderdale, Florida, U.S.A.

©1988, 1971 by Mark-Age, Inc.
All rights reserved. Published simultaneously in Canada.

This book, or parts thereof, may not be reproduced
in any form without the prior written consent of:
Mark-Age, Inc.
P.O. Box 290368
Ft. Lauderdale, Florida 33329, U.S.A.

ISBN 0-912322-06-3
Library of Congress Catalog Card Number: 88-61526

FIRST EDITION—1971
SECOND EDITION—1988

Manufactured in the United States of America
by Arcata Graphics, Kingsport, Tennessee

CONTENTS

MARK-AGE   1

INTRODUCTION   7

1. BIBLICAL RECORD   9
Abel And Cain—Golden Giants—Elder Race And Human Subrace—Abel And Cain Conflict—Awaiting Sananda—Spiritual Processing—Lemuria and Atlantis—Return Of Sananda

2. PHYSICAL BODY   18
Manifestation Of Light Bodies—Creation Of Elements—Creation Of Man's Physical Body—Relationship Of Man To Elements—Evolving Out Of Third Dimension—Coordination Of Light Forms—Erasing Earth Karma—Transmutation Of Physical Body—Balancing All Earth Forms

3. LIGHT BODY   27
Transformation By High Self—Self-Centeredness Caused Entrapment—Overshadowing By I Am Presence—Man Is Part Of Everything—Individualization Of Consciousness—Unification Of All—I Am Consciousness—Freeing The Entrapped—All Earth Matter Evolving—Freeing The Light Workers

4. ENVIRONMENTAL RELATIONSHIPS   34
Guidance From Higher Planes—Controlling The Environment—Elements As Part Of Man—Thought Patterns In Elements—Man And Those Thought Patterns—Mastery Of Self—Change By Thought—Transmutation Of Thought Patterns—Cleansing Of Soul Patterns—Impact Of Thought Patterns—Man On Earth Must Change It

5. FALL OF MAN   44
From The Central Sun—Experiencing Of Form—Experimenting With Life Form—Man Not From Lower Kingdoms—Final Fall—Memory Recall For All—Cause And Effect—Cleansing Of Error—Lemuria And Atlantis—Reevolving—Collective Society Necessary—Hierarchies—Transformation Of Earth

6. SONSHIP WITH GOD   59
All Is Of Spirit Substance—Manifestations Of God's Word—Concern Is With Present—Responsibility To Lower Kingdoms—Interplane Communion A Necessity—Light-Body Consciousness—Transmute And Transcend

7. FREE WILL　　67
Unique Function Of Light Body—Free Will In Evolution—Hierarchal Board And Free Will—No Free Will In Lower Kingdoms—The Seven Attributes Of God—Devic Kingdoms—Age For Revelations—Application Of Divine Attributes—End of Grace Period—The 144,000

8. DEVOLUTION　　76
Mastery Over Elements—Subconscious Track—Astral Areas—Devolution As Example—The Descent—The Entrapment—The Four Elements—Earth-Astral Cycle—Accumulating Knowledge For Mastery—Unification Of Individualized Souls—Collective Overcoming

9. SOUL　　87
Soul As Record Keeper—Soul Counterpart—Records Of Relationships—Devic Contacts—Astral Influences—Importance Of Soul Record—Responsibility In Freewill Choices—Soul Substance—Spirit Essence Contacts—Cleansing Soul Records—Projecting Prior Incarnations—Union With Soul

10. GOOD AND EVIL　　98
Man Created Evil—Man Against God's Purposes—Creating And Evolving Of Error—Destruction And Evolution—Re-formation Of Error Force—The Godlike—God Creates Only Good—Evil In Pursuit Of Man—Caution In Cleansing Error—Demonstrate Only For Highest Good—Evil Contrary To Divine Plan—Lesson From Evil

11. BRAIN　　108
Control By Spiritual Self—Improper Manipulations—Scientific Study—Bodies Of Man—Mind Capabilities—Relationships

12. DEVELOPMENT OF BODIES　　113
Sun Temple Records—Expansion Of Consciousness—Need For Soul Aspect—Creation Of Eve Or Soul—Rising Above Soul—Man's Seven Bodies—Individual Growth—Race Growth—Control Of Influences—Major Evolutionary Cycles—Understanding For Action—Use Of Information

13. REEVOLUTION　　125
I Am Self—Creation Of Man's Lower Bodies—One With Matter—Societies—Reascension Plan—Exploration—Fear Not But Seek Higher—Parable Of Adam And Eve—Parable Of Abel And Cain—Karmic Board—Lemuria And Atlantis—I Am Broadcasts—Ending An Evolutionary Experiment—Oneness With I Am Of All—Follow Spirit's Will

CHART OF THE SEVEN DIVINE ATTRIBUTES　　139

MAJOR EVOLUTIONARY CYCLES　　141

GLOSSARY OF NAMES AND NEW AGE TERMS　　145

EVOLUTION
OF MAN

MARK-AGE

MARK AGE PERIOD

The Mark Age period is that era known in prophecies as the latter days, the harvest time, the cleansing or purification period and the War of Armageddon. It is the time promised when there will be signs in the sky or marks of the age to alert man on Earth that he is indeed in those times when there will be the ending of an age, but not of the physical world. Having begun officially on Earth in 1960, the Mark Age period is scheduled to last the approximate spiritually symbolic period of forty years, long designated in religious scriptures as the period of fasting or cleansing the old in order to prepare for the new.

Thus Mark Age is that transition period of no more than forty years, and hopefully of considerably fewer, that is ushering out the old or Piscean Age and bringing in the New or Aquarian Age, the Golden Age when man will live in love and peace in a spiritual state, both individually and collectively. Actually, the Mark Age period is the final segment of a twenty-six thousand year plan and program by the Hierarchal Board, which is the spiritual government of this solar system, to bring man of Earth into the fourth dimensional or spiritual state of awareness.

MARK AGE PROGRAM

The Mark Age program is part of the spiritual or hierarchal plan and program whereby man of Earth is being informed and educated as to the meaning of this transition or Mark Age period. This has to do with his awakening to his spiritual identity, nature, heritage, powers and future, and to bringing forth on Earth a truly spiritual life and society in manners and means not even conceived as yet by most on Earth.

Man is an eternal being. Only by knowing of his true nature and potential, and of his past and future, will he be able to make the necessary decisions and to take the proper steps to enable him to achieve this change-over in the easiest and most desirable manner. For these changes will occur, regardless of man's desires and actions, since the entire Earth and all on it are involved. The Mark Age program is concerned with far more than aiding man to prepare for the New Age.

Mark-Age (hyphenated) designates one of numerous focal points on Earth for the spiritual forces from higher realms which are working in the Mark Age period and program. Established on Earth in 1960, Mark-Age since then has been publishing information and education in continual periodical form pertaining to this spiritual program. *MAPP* to Aquarius: *Mark Age Period & Program* contains some of the highlights from those publications, which have been read or have been heard to some degree by millions in many countries throughout the world. Although these publications have included varied types of material, most of it has been of the channeled communications through Yolanda of the Sun.

CHANNELING BY NADA-YOLANDA

Yolanda of the Sun is the spiritual name for this time for the primary channel of the Mark Age period and program and of the Mark-Age unit. She is one of many channels or interdimensional communicators who have served man for one or more lifetimes throughout eons. There now are many such channels on Earth performing various complementary and supplementary roles within the one and only spiritual plan and program called Mark Age by many but known also under different names. Known by millions on Earth as Nada of the Karmic Board, she is a teacher of channels and is on Earth now to teach and to demonstrate man's spiritual nature and powers and to help prepare him for his own spiritual awakening and his next step in spiritual evolution.

Yolanda is a conscious channel. She does not go into trance when in communication with spiritual beings of higher evolvements. Although trance channeling is a valid form, Yolanda never has done such in this lifetime. Prior to June 1960 her channeling was through automatic writing. Since then, except for several other instances in 1960, most of it has been speaking the words aloud as she receives them through mental telepathic impression or electromagnetic beam from higher dimensions and from interdimensional and interplanetary spacecraft. Occasionally some of the communications are typewritten directly as she receives them.

With but few exceptions, the vocal communications have been

tape-recorded as they were delivered, then transcribed by her. The original tape recordings still remain in the Mark-Age Library of Truth. The only changes made for publication are those concerning transposition of spoken to written language. All editing has been done by the Mark-Age editorial staff.

The interdimensional and interplanetary communications channeled through Yolanda and published by Mark-Age come from ascended beings in the celestial or angelic, the etheric or Christ, the higher astral or spirit, and the Earth or physical realms and from those of other planets in and beyond this solar system. Their sole purpose in delivering such communications is to help all on Earth understand and work for the spiritual awakening of man and the upliftment of Earth and everything upon it into the fourth dimension. In each instance since 1960, initial publication and distribution occurred within weeks or several months of the delivery date.

ORIGIN OF MARK-AGE

Mark-Age is not derived from any other organization on Earth. For neither Nada-Yolanda (executive director) nor El Morya-Mark (co-founder, who made the transition to etheric realms on April 24, 1981) had been members of any group, and had not studied any channeling or ideas, having to do with the spiritual Hierarchy. Thus none of the channelings through Yolanda or the other works of Mark-Age have been influenced in any way by what others have brought forth or have conceived or have perpetuated from prior works; although much of the hierarchal information and education released through Mark-Age supplements, complements, explains, enhances or corrects that of numerous other channels and groups.

Yolanda as Nada and Mark as El Morya, in the high Self states, have worked through many others, even having helped to found some groups, but in their Earth incarnations in the Mark Age period they did not study such works to any appreciable extent and were not members of, or did not have any influence in, those groups. Nor do they endorse and validate all of the past and current channelings and other information allegedly by or pertaining to them and other Hierarchal Board members, as released by some groups and individuals.

The name Mark-Age was revealed to Mark at Easter 1949 but no spiritual significance was given until 1955. It was not until early 1960, after meditations with Yolanda and participating in interdimensional contacts through her channeling, that the true nature and purpose of Mark-Age began to be given. Its start as a spiritual organization for Spirit's and the Hierarchal Board's purpose began with formal dedication and inauguration by Yolanda and Mark on February 21, 1960 in Miami, Florida. Since that time Mark-Age has functioned as a full-time spiritual unit of operation.

NATURE OF MARK-AGE

Mark-Age, Inc. is the legal entity for public functions of the Mark-Age Unit. Created in December 1961, it was chartered by the State of Florida on March 27, 1962 and was granted federal tax exemption as a religious organization on December 20, 1963. It is a nonsectarian, spiritual-scientific-educational society. All income prior to 1970 was through freewill or love-offering, tax-deductible contributions, which will remain one of the primary sources of support for this interdimensional, interplanetary and worldwide spiritual program. The desire and the goal are to teach and to demonstrate the Mark Age program to all on Earth, as well as to perform numerous other hierarchal plan works, for which purpose aid is welcomed from those who wish to assist and to participate.

Mark-Age is known as Unit #7 in its functioning on Earth for the Hierarchal Board. One of those functions is as coordination unit, not only between the Hierarchal Board and its Earth plane manifestation, action and workers but also for all hierarchal units on Earth. The symbol of Mark-Age is a seven in a circle. This has various spiritual significances. Specifically for Mark-Age, the seven denotes full Christ consciousness for the masses and the circle means this will be permanent or eternally closed.

Seven is the spiritual number for creation: seven rays of life or groupings of aspects and functions of God, seven Elohim in the Godhead which have created man and those of other realms, seven steps for creation, and seven bodies of man pertaining to his physical life on Earth. Mark-Age as Unit #7 also indicates it is the main focal point on Earth for Sananda and Nada, co-Chohans of the Seventh Ray. The Mark-Age theme and motto is *Love In Action*. This is the Christ principle and function. The New Age is action with high Self, action with love.

As an externalization of the Hierarchal Board on Earth, Mark-Age has numerous purposes and functions. The primary one is to teach and to demonstrate the Second Coming of the Christ. This has two meanings: the return of Sananda, spiritual ruler of Earth, as Christ Jesus of Nazareth, and the return of man of Earth to his own Christ consciousness and spiritual powers. Another major function is to implant spiritual government on Earth.

DIVISIONS OF MARK-AGE

There are five major divisions of Mark-Age for the externalization of the Hierarchal Board upon Earth. These five, in operation and to be expanded continually, are *Mark-Age Inform-Nations (MAIN), Mark-Age Meditations (MAM), University of Life, Healing Haven,* and *Centers of Light.*

Mark-Age Inform-Nations (MAIN) disseminates the Hierarchal Board's desires, plans, news, information and expressions, via *MAIN* news services, informational material, periodicals, books, recordings of every nature, public appearances, and every other method of communication. It is the main route through which the Hierarchy teaches and demonstrates the educational program for the spiritual awakening and development of man on Earth, through the Mark-Age teachings, information and functions.

Mark-Age Meditations (MAM) is an interdimensional and international network of spiritual groups and individuals coordinated for helping to bring about the Second Coming and other hierarchal purposes. Established in May 1972, it originally consisted of the playing of specially prepared weekly broadcasts on tape recordings as a prelude to meditation by Mark-Age groups and members and others. Each *MAM* focus works on the same spiritual project during the same week, thus providing tremendous spiritual energy for manifestation of specified programs. There also are other types of *MAM* programs for varied interests and functions.

Healing Haven is the Hierarchal Board's method for presenting the best of information, techniques and research for the spiritual healing of man's four lower bodies (physical, astral, mental and emotional) and his mortal consciousness (conscious and subconscious or soul aspects). It embodies the best of past and present methods for healing the whole person and explores and offers new avenues, all to aid in eliminating man's dis-ease and suffering and to help bring him into balanced spiritual expression.

Centers of Light are focuses of two or more individuals coming together for learning and expressing spiritual knowledge so as to help themselves, their fellowmen and the hierarchal program. The *Centers of Light* include not only Mark-Age, *MAM,* and *University of Life* groups but also Mark-Age-affiliated and independent units of spiritual light workers. Mark-Age Unit #7 is the prototype of such centers, offering valuable experience and assistance.

University of Life teaches and demonstrates spiritual living, learning, expression, and growth of consciousness. Its basic structure is the twelve schools of Mark-Age. It is not a formalized system of education in any one place or manner. It does not take the place of or supersede formal educational systems or institutions, but enhances, supplements and complements them. The home, school or other environment is the classroom. There are no degrees offered. The courses, the rate of study and the benefits are determined by the *Life* student alone. Mastership of life is the purpose of *University of Life* study.

SCHOOLS OF UNIVERSITY OF LIFE

(1) *Being:* study of the nature of Spirit, with development and use of Christ consciousness and powers by the individual. (2) *Education:* spiritual educational programs and systems. (3) *Geophysics:* study, control and development of land, sea and space. (4) *Government:* study of the spiritual, inner plane government of our solar system and how to pattern local, state, national and international governments on the same principles, ideas and experiences. (5) *Guidance:* assistance in determining courses of action for individuals, organizations and governments, based on spiritual laws. (6) *Health:* teaching eternally perfect health of mind, body and soul, and how to attain and to maintain it. (7) *Living:* increasing the enjoyment of abundant spiritual living through entertainment, recreation, fine arts and other means. (8) *Love:* instruction and demonstration in nature and use of divine love in personal relationships. (9) *Protection:* the invocation of divine protection and how to secure it for one's self and others. (10) *Religion:* practical metaphysical interpretation of religions, in a nonsectarian manner. (11) *Science:* bringing forth of spiritual-scientific principles and products. (12) *Supply:* teaching and demonstrating man's ability to control material needs and products, both natural and manufactured.

INTRODUCTION

The Holy Bible contains the story of the creation of this solar system and man's inhabitation of it. As is true also of many scriptures and inspired revelations, much of the nature and the history of this creation and of the evolution, history and spiritual development of man is presented in parable and allegorical form. It requires special illumination, by either communication with those in higher realms or contact with one's own high Self, to interpret these codes.

Both forms of illumination brought forth the information revealed and recorded in this book. Chapter 1 is mainly the result of Yolanda's intunements with her Nada high Self gained while driving with Mark through much of the United States in late 1966. It is based on material given in *New JerUSAlem,* published by Mark-Age in 1967. The Salinas Valley intunement and the discourse by Lord Uriel were received by Yolanda while driving through that area with Mark on July 31, 1970.

Chapters 2–13 are the discourses channeled through Nada-Yolanda by Hierarchal Board members Lord Uriel, Archangel of the Seventh Ray; Sananda, Chohan of the Seventh Ray; and Nada, Co-Chohan of the Seventh Ray. Chapter 2 was delivered on July 29, 1970 at tiny Lake Madrone near the village of Berry Creek, California, about twenty miles northeast of Oroville. Chapters 3–13 were channeled in Santa Barbara, California in August 1970: #3—12th, #4—14th, #5—16th, #6 & #7—17th, #8 & #9 & #10—18th, #11—20th, #12 & #13—21st. The locations are given because they are within or near key areas mentioned in the discourses.

The glossary contains explanations of possibly unfamiliar terms. But it cannot even begin to give the comprehensive information and education concerning the topics mentioned in this book, and many more, pertaining to the hierarchal plan and program for the spiritual awakening and evolution of man. The basic text and reference that does provide this material, plus a glossary of two hundred and sixty-seven New Age terms and a twenty-two page index, is *MAPP* to Aquarius: *Mark Age Period & Program.*

The chart of the seven divine attributes provides a reference to those attributes as discussed in the text and gives their relationship to the Seven Rays of Life and their application by man on Earth. The summary of major time cycles presents an expanded outline of those periods as explained in this book.

1. BIBLICAL RECORD

ABEL AND CAIN

Genesis to Revelation in the Holy Bible is an outline of man's evolution prior to the fall of part of the race into Earth incarnation, of the subsequent devolution, of the reevolution to the true status as sons of God, of prophecies concerning the latter days of the two hundred and six million years of evolutionary experimenting on Earth, and indications of the next level of evolution beginning by the end of the twentieth century.

Man of Earth operates on three levels: Christ or etheric, operating in his light body in the fourth and higher dimensions; Adam or mortal, striving in his physical body on the third dimension of the Earth plane; and Eve or soul, functioning in his astral or emotional body in the astral or psychic planes. When Adam (conscious) and Eve (subconscious) dwell together on Earth (in the physical body), they bring forth the children Abel and Cain (thoughts and deeds of good and evil). Thus, the allegorical story of Adam and Eve and of Cain and Abel is partly a spiritual and a metaphysical teaching concerning man's own nature and creations.

But also it is historical in that it is based on the long struggle between two factions of the Elder race, the golden giants spoken of in the Holy Bible and other scriptures. The conflict between the Abel and the Cain groups of fourth dimensional man concerned their responsibility toward that part of the race which had descended, or had fallen or had been entrapped, into the third dimensional or physical matter of Earth. This third dimensional segment of man had devolved into the human or the subman form to use the Earth plane as a schoolroom for experimentation and exploration. Becoming entrapped in matter, it lost the use of spiritual and psychic senses and became confined to experiencing only through the five physical senses.

The Abels, particularly their leader Sananda, desired to teach their human brothers the powers and uses of the light or fourth dimensional body so as to raise them into the spiritual consciousness of the Elder race from which they had fallen. But the Cain clan was of the opinion that those in the third dimension had been placed on Earth to serve the Els, to be slaves. Since the life of the Cains was so pleasant and easy, they believed God preferred them, and that all they thought and did was favorable to Him.

GOLDEN GIANTS

The Elder brothers served as rulers and priests over man in the third dimension of the Earth plane. In their etheric or light or fourth dimensional bodies, the Els could travel interdimensionally and interplanetarily throughout the solar system. They had memory recall of their past-life experiences and places of existence in the solar system. They were able to maintain the same body or vehicle in which to express on Earth for at least two hundred and fifty years.

Their social structure was truly communistic, in that no one owned anything. But their society was a class system, separated according to spiritual evolution and talents. The seat of one of their great tropical civilizations was a planned society developed around an inland sea now known as Salt Lake and the Salt Lake desert. Their settlements were built around the sea in circles or tiers, for grading or segregating purposes. The higher mountain ranges were set aside for spiritual retreats, schools, and training centers for priests. There the golden giants, the sons of God, grew in self-esteem, self-satisfaction and conceit.

The governmental capital of the Elders was in the area now known as Lake Tahoe in California and Nevada. The capital was built in concentric rows, according to their favorite custom. This was most suitable, due to the high ranges which completely closed it off and prevented the humans from entering the seat of government. This physical inaccessibility in no way hampered the Elders, who functioned in the fourth dimensional body. They could enter or leave either physically or in the light body, since they had control over both third and fourth dimensional frequencies.

ELDER RACE AND HUMAN SUBRACE

The third dimensional or human race operated in vehicles quite similar in structure to our present physical bodies. They were able to maintain those bodies for seventy to eighty-five years, a few living to be about one hundred. They understood most of the teachings of the Elder race and benefitted from the interdimensional and interplanetary travels and exchanges the golden giants enjoyed. But they were prohibited from participating in the inner temple practices and the esoteric mysteries. The higher ranges, both literally and symbolically, were off limits to them.

Eventually, some of the Elders believed they could use, command, even suppress, the further development and spiritual growth of the human group. Inevitably this trend led those of the Cains to think of the humans as less important in the one brotherhood of man under the Father-Mother God. So, instead of developing the intended teacher-student relationship to bring about the balanced third-with-fourth dimensional form which had been the original intent and purpose of the guardians composing the Hierarchal Board or spiritual government of our solar system, the situation became more separative and restrictive in nature and the relationship became that of masters and slaves.

When the council of the fourth dimensional race determined to eliminate any representation or consideration of those on the third dimensional frequency vibration, Sananda championed the concept that the Elder race was responsible for the human race and was supposed to teach the humans higher and proper uses of their inherent spiritual nature and powers, otherwise the latter could not evolve from the density of Earth plane matter. Arguments and strategies took possibly thousands of years, developing through one or more incarnations and societies toward the actual planning and execution. But the present Lake Tahoe area is where policies were determined and decisions were enforced.

In this argumentation, Sananda was not one of the governmental leaders but was at the spiritual headquarters of this race of golden giants. This was located in what are now the Grand Tetons of Wyoming, covering the Yellowstone Park area. It was the approximate center of their spiritual retreats and schools. Like the governmental capital, it was built in concentric circles encircling a desert and an inland sea. The main source of light and energy for the Elders at that time was the Temple of Venus, located in these mountains. To this day those who are spiritually intuned can feel the special radiations of brotherly love and mortal sacrifice that emanate from this focal point.

ABEL AND CAIN CONFLICT

More and more the Abels and the Cains became separated in viewpoint and purpose, until finally some of the Abels, under Sananda's leadership, determined to lead an expedition across the inland sea covering what now are Utah and parts of Nevada. They moved to the northwest, crossing the Cascade Range in northern Washington, through what is now British Columbia to what we know as Alaska, which then had a temperate to semitropical climate.

The last battle between the two dissenting forces took place near the foothills of Mount Shasta, about two hundred miles northwest of Lake Tahoe. The Cains, in the majority, tried to prevent Sananda and his band of disciples and a small group of humans from leaving their control and rulership. This so-called war was fought on a higher level than the physical as we understand it. The battle was one of spiritual talents and wills. Some of the records of this titanic struggle have been transmuted into crystals of an etheric vibration and now are locked in the third dimensional frequency inside caves or tunnels around Mount Shasta in northern California.

It was here that Sananda met and passed his great test of that period. Through divine love he discovered the key in the vibration and mantrum known as *Love God and Love One Another.* Use of this key raised the crystals and hid them beyond the selfishness and the spiritual pride and ego of those who did not want this racial history preserved or made available to future generations. This initiation gave Sananda the foothold and the authority thenceforth to preside over the Seventh Ray of Divine Love, Peace and Rest for all third dimensional mankind during the dispensation which followed. Eventually it led to his Hierarchal Board appointment as Prince or spiritual ruler of Earth.

Yet, in spite of having passed through this grave trial, Sananda as leader of the Abels was overpowered; Cain slew Abel. This prevented him from taking command then and there over the future evolvement of the third dimensional race on Earth. The human subrace thus lost its opportunity for direct, in-person training in a peaceful, constructive, brotherly pattern.

The Hierarchal Board then banished all the golden giants to Venus, home of the Third Ray functions and aspects in our solar system, for their further cleansings and greater dedications. Although they had made mistakes and had lost a golden opportunity to express greater love for a less evolved race, they were not lost or punished in the mortal sense of the term.

AWAITING SANANDA

This war of the golden giants is perpetuated in our racial memory through the many allegories, myths, legends and folk tales to be found in nearly all cultures and societies. Amongst the American

Indians, for example, are many stories relating to this, including the coming of man from the skies to visit Earth. Tribes in the north and the west still cherish secret places of worship which commemorate the home and the temples of Earth's spiritual leader: Sananda, the golden one. Through their arts and crafts they still pass down to present generations the story of his struggles for mankind. These appear in their tribal dances, hieroglyphic tablets, tapestry and pottery. But only a few of today's Indians have been initiated in this knowledge.

The Salinas Valley, near the coast of central California north of Santa Barbara, is very holy ground. It has been protected by Indian and devic forces for the return of Sananda. It was there that the Abels had been able to find some peace and harmony, with a small segment of them solidifying their resources and implanting a number of vibratory rates for eventual resettlement.

The history there is confused because some wished to pursue their goals, whereas others decided to retreat and to declare a truce with the Cains, letting the higher forces determine balance in ages to come. But nothing was resolved except the implanting of a high rate of energy, which since that time has been felt strongly by those sensitive to higher frequency rates.

It has been a favorite base of operations for etheric spacecraft. Even during the days of the AmerIndians here, huge spacecraft would land from time to time to allow conversing with the chiefs and the holy men. This did not occur at frequent intervals, but perhaps every two hundred and fifty to three hundred years. Enough source of interest was secured from these contacts to cause legends and the beginning of communication between those in the higher planes and the remnants left on Earth.

It is not easy to convey the importance of these contacts. They are part of the divine project and system of never completing a segment until it has been fulfilled via spiritual enjoyment and satisfaction. No two systems are alike. No two plans follow the same course of action. Only are there similarities and patterns projected in order to show a divine pattern of procedure in the upliftment or evolution of any specific job well done via divine fiat.

SPIRITUAL PROCESSING

"Uriel. Let us begin to know this sameness of pattern so we can begin to detect the schematic of spiritual processings. All are one in the Divine, and no two parts of the One can ever be separated. Therefore, no matter how highly developed a segment is, it can never lose touch or communication with another part of the whole or the one divine object.

"Let us know, therefore, that those called the Elder race are no more separable from man of Earth. But they take on different form. They know different jobs and responsibilities and they know they are obligated to their function until all are raised to the level within the same specific species or evolutionary growth pattern. This is not easy to explain, because so much time is involved in the process that man thus loses touch with all the various parts of the whole plan and the oneness which is God.

"As is indicated here, many are on Earth from the Elder race or White Brotherhood and have not fallen into the trap of three-dimensional feeling or ensnarement. But they still can take bodily form and not stand out amongst those of Earth because their nature is so similar to those in that revolting trap of three-dimensional cyclic development.

"However, every time they take on body and work within the mold of this segment of the one race, the sons of God, they change, reset, formulate new concepts and precipitate a higher action within man's racial patterns and thinking. This takes billions of years. Since man is in no rush to reach his spiritual reevolution, it is slow in turning around from mortal into immortal intunements.

"He will have to see gradually the benefits of operating spiritual vehicles at his disposal before he is willing to overcome his mortal aspects which he himself created along with devic and elemental kingdoms at his disposal and under his command. I am Uriel, lord of light and love divine. Amen."

LEMURIA AND ATLANTIS

Regrettably, the destructive motives, thoughts and acts of the Cains greatly lowered the standard or level of frequency for the Earth. But despite this they were able to maintain firm control over the planet through a series of civilizations. Among the more recent were Lemuria or Mu, in what is now mostly the Pacific Ocean and the Pacific coast of the United States, and Atlantis, from the eastern coast of the United States across the Atlantic Ocean to the western coast of Europe.

The height of their successes and developments was approximately twenty-six thousand years ago; the final and last days of their destruction were twelve to thirteen thousand years ago, which is about as far back as we have had acceptable physical documentation. Still, their approach was not based on the divine law and truth of *Love God and Love One Another*. Instead, they continued to use force, domination, control and various types of dictatorships against one another. Through such designs of destruction, the law of cause and effect brought them and their civilizations to yet another cleansing, referred to as the great flood of Noah's time.

In the hundreds of years it took for the final destruction of Lemuria, the lower grades of workers in that civilization took the seed or the pattern of their bodily forms, by means of migrations, to Alaska, the northern United States, and Canada, with many of them escaping to the islands of the Pacific which then were the mountain peaks of that sinking continent. The higher grades of fourth dimensional form no longer were operable on this planet. And until the return of Sananda to Earth, they will not be operable for the majority of mankind here now. He personally will teach greater compassion and love in the Golden Era, beginning about 2000 A.D. By then the memory scar of this advanced but degenerate society should be healed, with all spiritual conceit and pride removed from the subconscious tracks of the race.

It may seem that the transmutation of mortal into spiritual consciousness has been long and difficult. But what the Elders did achieve during the age of the golden giants was an anchoring for a future racial pattern. They succeeded in leaving their imprint on the third dimensional frequency of this planet, even though it has taken the masses of mankind millions of years of physical and spiritual evolution to rise to another peak of development that will raise the third dimensional form into fourth dimensional frequency and expression.

During the times of Lemuria and Atlantis the spiritual Hierarchy came to the sad realization that the human race on Earth still was not ready to complete its final, seventh step into fourth dimensional or spiritual consciousness. So, a new plan and strategy were conceived. Presently we are in those latter days of the greater, overall, longer ranged, two hundred and six million year program. This Mark Age period and program therefore are the last judgment spoken of in the Holy Bible.

RETURN OF SANANDA

Each one's individualized Christ or I Am presence and the collective spiritual organization headed by the Hierarchy of this solar system force all souls, and the race itself, to return and to work out their karma. That which is sown must be reaped. So it is that

Sananda, the Christ leader of the Abels, not only has earned the privilege but is obliged to lead mankind of Earth into spiritual evolution. Continually, in various roles and incarnations, he has returned to guide the humans out of their bondage in this dense, material, third dimension.

However, from the time the Hierarchy banished the golden giants to Venus, and even through the thousands of years of Lemuria's great civilization, Sananda did not take up physical incarnation on Earth. But this did not prevent him from guiding, teaching and instructing man from the inner planes in much the same manner he has been doing in these latter days, the Mark Age period.

In the last days of Atlantis, Sananda did incarnate on Earth in order to guide the cleansing and to preserve a certain amount of balance, understanding and perseverance within the remnant of that civilization. This return was in the generations of Noah. Noah was not an individual, but was that group of men on Earth which did the will of God. Through their spiritual knowledge and experience they managed to preserve one male and one female of each species. This is symbolic reference to their ability to maintain the proper balance of positive and negative polarities needed to perpetuate each type of life form on the physical plane. The leader of the Noahs was Sananda.

The last thirty or forty years of the twentieth century are the latter days. This end time is not the end of the world but is the final judgment or the final campaign of the most recent, a twenty-six thousand year, hierarchal plan and program to raise all on Earth into the fourth dimension. All that the sons of God have worked for in this solar system for the past two hundred and six million years is about to culminate, preparing the way for a new cycle of evolution. But one of the major tasks remaining in this Mark Age or latter-day period is that of preparing the way for the return to Earth of its spiritual ruler, Sananda; who, in his last Earth incarnation, was Jesus of Nazareth.

Spiritual evolution occurs in orderly, spiraling stages. At the peak period of each age in the evolutionary struggle of man on this planet, Sananda as Prince of Earth has reappeared to lift, to instruct and to help in the gradual transmutation of man's expression on this third dimensional frequency. Each of his incarnations, for each period of evolution of man on Earth, initiated and patterned another firm step in the ladder of spiritual growth. Some of Sananda's better-known incarnations on Earth are given in the glossary of this book.

Finally, as Jesus of Nazareth, he fulfilled the prophecies that are found in almost all religious scriptures concerning the anointed one or the fully spiritualized man on the third dimension in the higher body, the light or resurrected Christ form. By transmuting his third dimensional body into his fourth dimensional or light body within this Earth frequency vibration, this way shower for man on this plane dramatized the pattern for all to follow in the return to being fully functioning members of the Elder race.

2. PHYSICAL BODY

MANIFESTATION OF LIGHT BODIES

Dearly loved children of Earth, I am here amongst you as the lord and overseer of devic forces in the Earth planet. According to your terminology and understanding, you hail me in the name of Lord Uriel, Archangel of the Seventh Ray of love, of peace and the joy in divine love for the Creative Force in all natural influences.

You will begin a new phase of operation. You will begin to understand many things referred to as the coordination of light frequency in and between Earthlings and in and between Earth formation and higher plane or inner plane manifestation. You have begun to seek the many things that are natural in your environment, to aid you in this ongoing and in this cooperative venture of life form with life form. You will be given detailed instructions in the next few months of your time to help you in creating the light frequency in yourselves and in those who are aligned to your inner plane development and your outer plane manifestation.

In all these things we are requiring and seeking a highway into which you may achieve, and consciously work through from, mortal into immortal or light-body manipulation and manifestation. You will begin to see the light-form around all nature and natural forms of Earth plane living.

You will begin to see and to know your own light form or etheric substance as it projects from you and projects into you in the Earth plane embodiment so you may be lifted and transmuted and may seek the resurrection of the spiritual Self in outer formulation. That has to be the entire substance of your concern and of your interest and of your development at this period of time in the sequence of hierarchal unfoldment for Earth planet and the souls of men now incarnated upon the Earth planet, before going into the Golden Era.

You have been given much to understand all along these lines before this broadcast of light energy, so it is understood that much frequency evolvement has taken place through hieronics broadcasts already, and through the space manipulation of many craft which are anchored above your Earth plane atmosphere in order to aid you in this ongoing and upliftment.

But until you have absorbed and have attributed all of these knowledgeable facts, we do not anticipate that you can accept and comprehend every little individual piece of information and development adequately in the conscious and subconscious realms of your own body and mental responses. Kindly give this a little respect in that aspect: that you know and realize all has not been absorbed sufficiently for you to be able to grasp every detail I speak of and that is now being unfolded unto Earth plane consciousness.

CREATION OF ELEMENTS

I give further information regarding this particular broadcast at this particular site, for a specific reason. It is because you are in the approximate area of recorded history and in the geographic location whereby you can tap in to prior history, according to Earth plane matters, that will be extremely significant to Earthman evolvement and development from this moment onward. I refer you to historic periods of Earth plane experience prior to anything now in existence in a physical manner so far. Only through the channeled communications and the intuitive impulses of those who are sensitive enough and connected sufficiently with the past recorded history of mankind in the subconscious realms can you now comprehend what I will bring sufficiently to the forefront. [See page 7.]

In eons past, when man's body was being formulated out of the etheric and elemental substances for this particular plane and planet, we had a contrary civilization of inner plane and outer plane developments, working in conscious control, resting their history and evolution upon the manner in which they would respond to one another. That era was twenty-six million years ago, in its entire form. It expressed as far back as two hundred and six million years ago in being resolved for that particular episode.

In this era of which I speak, prior to any known knowledge historically or scientifically as you might place it, my own forces were experienced in taking the natural elements sufficiently responding to this age, to this era and to this place of Earth within the solar system evolutionary cycle. We worked upon the elemental force in order to bring into manifestation a number of forms which you recognize as the primary elements of your planet.

Out of this substance and out of these elements man was created, or given, a human form. His substance originally and primarily is of etheric nature or spirituality. A form of this nature exercises many privileges and symbols which are not understood yet on the plane or dimension of physical expression.

In the original state, man has not had physical form. But in the primitive age of which I speak, when this planet was being formed on the physical, we of the Hierarchy and those of us of the angelic realms saw fit to bring into form the elemental kingdom through the natural resources that were given to us in three-dimensional matter. It called upon much activity and knowledge and control, and seeks from it a type of consciousness prior to the spiritual evolution of the form, of the place and of the planet.

When this becomes necessary to our conscious evolvement and to our conscious dedication to a hierarchal plan, we can bring about many new levels of comprehension and of development which are required of us. This means we respond by our nature and by our function to that which will serve the overall good of a higher plan not known previously or not understood by all levels with which we work or all forms with which we work. In other words, what we do with the elements, from the archangel and angelic realms, are not known always or not conveyed always to the elemental kingdom.

CREATION OF MAN'S PHYSICAL BODY

Thus we made form, or brought forth form, through manipulation and control over the substance or the origin of these elemental kingdoms. Out of this we were permitted to make form into which man was allowed to experience. Man, in this sense, is to be designated as the spiritual man or mankind, the son or the soul from which he had come or had seen life here upon the Earth in his spiritual or etheric form.

So he became attracted to the physical form which we had conveyed and which we had manipulated in the three-dimensional developing stages of Earth in evolution throughout the relationship it has with the rest of the solar system.

It is sufficient here and at this time to say and to convey that man of himself did not evolve out of the elements, nor did the elements precede man, but that man as a spirit preceded all these things and participated in all these things after we had helped to prepare the kingdom in which he could have his reign and experience.

RELATIONSHIP OF MAN TO ELEMENTS

It is out of this era and time that man now finds his evolution in that form nearly completed. For he has experienced totally or has utilized totally all the power he is going to be permitted in the elemental life form. That is, he has utilized every part of his conscious control over the elements of the Earth planet as it now exists and as we had helped to create it into physical formation.

Now that the planet, as well as the solar system, is to evolve beyond this stage of physical development, man must be brought into conscious awareness of what he has, what he has come from and what he has been aware of for eons of time in this evolutionary process. Man has come full cycle around to the beginning stage of where he became attracted to the elemental era and origin of formation.

He has broken down all the elements to their original form and has learned to manipulate them and has learned to put them into new properties and to create new forms out of the elements themselves. This is as far as he must go in his realization of elemental relationship or elementary kingdoms within his own frequency of operation. Until he learns the balance of these things and the interrelationship of these things he cannot be given more power over them than he now has gained.

In addition to this, he must come to know and to recognize that this power he has over them is his spiritual power and not his physical power; and that he is not related to them in eternity according to the form he has taken or they have taken, but in the essence or in the spiritual vibratory frequency from which all things have been created.

He must come to know that his relationship to the basic elements is in his form with them on a physical, three-dimensional planet; and not that he has evolved from out of those forms or from out of those elements, but that he himself participates with them in the creative evolutionary process of form or matter in eternal substance, which is the etheric level of creation. It is a very individual existence that each form has and it is a very individual formula that each form has created out of the balance of one element or chemical essence with another so that varied forms and various functions can be performed and interrelated and balanced for a whole concept.

In these eras of *now* he will be given this knowledge. He will be given recall of these matters in quite dramatic instances. He will be able to go back to the beginning or the origin of this development and see where he has taken too much liberty or too much leeway in expressing himself and in experimenting with these various forms. That will come about in the next four to forty-four years of time upon the Earth planet. It will create much havoc in the balance you now have on Earth frequency or Earth development. It will create and will bring about a new consciousness within mankind himself.

EVOLVING OUT OF THIRD DIMENSION

One of the things that are most essential to learn and to understand in this present environment and place geographically is that where you are in the area of time and space was approximately where these experiments had begun to be made manifest upon the Earth. It is where the Elder race, in fourth dimensional frequency, had begun to allow third dimensional form or matter to come under its reign and obligation in order to bring it into subjugation and to teach third dimensional matter and form how to evolve into fourth dimensional frequency and form.

When this did not succeed according to the divine plan, man was trapped in the third dimensional frequency and had to evolve up out of it; and so has been evolving in this form for the last twenty-six million years, although the beginning of this experiment goes back to two hundred and six million years of approximate, recorded solar system history.

As you have been told, especially through this channel, all are going to be given enough information about the solar system history so he or she can relate to his or her own spiritual history and evolvement in this solar system as a soul in evolution, and in a soul who is about to graduate into a new form and a frequency cycle that shall allow totally new experiences and new evolutionary developments in the race of mankind within the entire solar system projections. Under these auspices come I, in my particular function regarding a happy and fulfilling relationship with devic elements and all who are consciously working toward light-body projections.

In the days of these many eons past, the light body was the form in which the Elder race did operate in this approximate geographic area. Of course, it takes in many thousands of miles of area, and we are not pinpointing this to one specific place. But here or nearby were many events that, geographically located, could be related to the evolvement and decisions of an Elder race in fourth dimensional frequency body aiding and assisting third dimensional matter in evolutionary growth. Therefore, it is approximately appropriate to begin again in expressing, and showing forth, light-body form and evolvement here amongst the same elements where the struggle for this had taken place.

It is essential that man begins to realize that his present limitations are derived from those errors and mistakes of his entire grace or gradual growth as a race, and not necessarily as an individual. It is important that mankind recognizes that each element within himself is to be raised, in order to project the light-body form. It is essential to know that the transmutation over the elements must come via the light projection of his spiritual Self so that, as an individual, man can raise the elements of his body via his own light-body projections and by his own consciousness as a spiritual being rather than as a physical being.

It is important that man realizes that each individual person who so achieves this can aid the entire race and the entire planet; for as he raises his own elements he raises the elements everywhere upon the planet and thus he affects all life form and his fellowmen equally and as well.

COORDINATION OF LIGHT FORMS

All this is to be brought into conscious awareness by the light workers on the inner plane and on the outer manifestation of life here upon the Earth. It is essential that this be done shortly, for we are about to begin a coordination of light centers and light workers in a manner which we never have achieved before in the present cycle of historical events upon the planet.

Not since the days prior to what has been mentioned here, the civilizations of Lemuria and Atlantis, have we been able to achieve any form of coordination between the inner and the outer, and all life form on the outer, simultaneously. This had been achieved in that time prior to Atlantis and Lemuria but has not been achieved since that time.

In the struggle for coordination and cooperation between all the light forms, as relating to the Earth planet and the Earth form upon the Earth planet, we have no precedent upon which we can base our achievements and our goals. Therefore, we expect your cooperation as you can give it, where you can give it and as much as you can give it.

ERASING EARTH KARMA

Many new light workers are being informed and awakened at this present time. Many new light workers, in the sense that they are very new upon the Earth planet, have not had much karma with the Earth planet and therefore have come in with comparatively clean records as far as the Earth planet is concerned.

This has been extremely necessary, for this reason. In view of the fact that the karma of Earth has been so deep and so overwhelmingly difficult on the negative side or the debit side of the ledger, we have found, in our evolved state of consciousness and in our Karmic Board discussions, that the best and highest way we can bring about a total swinging away of the Earth from these deeply negative situations is to overwhelm the planet with higher frequencies in the souls incarnated upon the Earth for a higher achievement.

We have invoked and have provoked many souls of higher evolution to come upon the planet in bodily form, and in etheric spacecraft if possible, to bombard the Earth with their higher frequency ideals and with their abilities, to achieve a greater demonstration than hitherto has been shown on the Earth in order to dissolve all the negativity and unpleasant race memory as much as possible while we go through this cleansing and move into the Golden Era.

This does not mean those souls can erase or eradicate the soul karma of any single individual other than themselves. It does not mean they can obliterate the karma of the Earth planet itself, or the race in connection with the Earth planet. But it does mean that, like fresh troops, they can come in and win a single battle in a long war to bring about fourth dimensional consciousness upon the Earth and thereby let those of Earth who are connected with it have a better battleground, so to speak, on which they can fight the war of achievement; achievement for their Christ Selves in manifestation through physical form, or physical form as it will be in the higher frequency of fourth dimensional matter.

All this is part of a divine hierarchal plan. All this has been prepared for, according to your recorded history, for the last twenty-six thousand years. In other words, I here state that those souls to which I am now referring have not incarnated frequently upon the Earth, perhaps, but would have to have enough grounding upon the Earth in order to achieve this job and function as I have outlined it here.

TRANSMUTATION OF PHYSICAL BODY

Now that I have set the scene and have created the essential overtones toward your responsibilities, I will give you much more news and information in the short time ahead of us, in the attempt to achieve exactly what I already have given briefly or have outlined for a moment in space and time here. You have been brought purposely unto this spot so you may have recall and you may have re-establishment with those forces that can guide and sustain you in the troubled times ahead.

You will be given much in the way of responsibility to do this. You will be given much inspiration. All of it will not come in a sudden moment. You will be expected to demonstrate many things, out of your conscious awareness of these events, and not know at the moment why it is necessary. You will be led from the higher Self and from the higher forces which are now connected with you through this.

As you can be made aware of the elements that are achieving this properly and have raised themselves sufficiently to a purified state of elemental existence, you will be able to draw in, and to respond to, those elements and thus raise and eliminate from yourself those needs for lower elementals in your own frequency.

This is one of the ways in which we raise the physical frequency that you are experiencing into the fourth dimensional or light-body frequency; because, as you come to understand, and to relate to, the lower elements in the natural kingdoms all around you on the Earth, you no longer have to experience them within your own orbit or frequency or body. You will rise above those particular elements in yourself and thereby will eliminate them from your natural expression in the vehicle.

You thus will transmute the physical vehicle and will be able to have your future existence in the fourth dimensional frequency body. This is the light body or the etheric substance, which is eternal. To express this in other Earth terms: it is the resurrected form which was demonstrated by the Master Jesus of Nazareth, who was the Christ of Earth and is the Prince of this particular planet, this form and this dimension, and over all these elements, in his Sonship with God.

But because you are part of the parcel or part of the form out of which he was created as a soul, you too can achieve the same resurrected form. You too can rise above the elements of the baser form, or the third dimensional frequency body, and transmute consciously your physical body and achieve consciousness in, and operate through, the light form or resurrected body and walk upon the Earth, walk through the Earth, conquer that which is of the baser elements or the lower elements to your higher Self function.

BALANCING ALL EARTH FORMS

This is all for the moment. This is your goal and your outline for the next immediate four-year cycle of time. This shall be your total concentration. Through this the elementals who are based on Earth shall be purified, shall be cleansed, shall be raised and shall be rebalanced as you rebalance your own consciousness.

As you achieve power over the elements in your own body, so you will achieve and will balance those in the other forms of Earth and will bring Earth back into its proper, total state of purity—the state of Eden, so-called—in which all are in proper form, all are in proper relationship, all are in proper harmony with one another; and the form of man, as a son of God, can have power over, and work with, all in love, peace and justice. That divine love, which I spoke of in the beginning, for all life form and energy will be achieved and will be exercised and will be executed by the Sonship, through God, of man by the year 2000. Amen.

I as Uriel guide and control many of these things so you may have assistance and perseverance. God is with us and in us, and for this I am grateful. Amen. Uriel.

3. LIGHT BODY

TRANSFORMATION BY HIGH SELF

My name is Nada. When I, in the light body, overshadow that which is my physical instrument and soul consciousness now known as Yolanda of the Sun, I evict all lesser energies and elements, while transforming much that subsists in the soul record and body of that consciousness. It may be termed, in some degrees, as the descent of the Holy Spirit as pronounced in other edicts and philosophies.

But should you examine this from a scientific point of view, you have to recognize that a higher force or a higher energy cannot exist in the same vessel as that of lower substance. For those moments and those times I, as the spirit of this individual, influence and succeed in eliminating all else but that which is pure. It is part of divine principle and right.

It is not so much that, once this contact is made and produced, all else is eliminated and gone, but that for frequent moments of such communion of higher Self with soul body and spiritual essence I can, or the spirit can, stimulate and produce a higher effect in and through that product which you call the body, the soul and the mind.

It is for this very purpose that such manifestations do occur. At the proper moment, when it is for the higher good of all who are related to this entity and substance matter, the production can be complete and the communion everlasting on this plane and development. But many fractions are involved here; not only the fraction of the soul and the mental level of the individual through whom I respond and work for a time, but the many fractions of relationships who are part and parcel of the experience.

SELF-CENTEREDNESS CAUSED ENTRAPMENT

This does include the entire race consciousness. For no one is separated from the entire race, but all are part of that race through this experience of I Am consciousness and mortal body and soul evolvement. Until man has come to realize that nothing is separated, either in the physical level or in the spiritual level, he never will progress beyond his subhuman nature and his animalistic reactions and his selfishness, which is animalistic. For it is only in the animal kingdom that self-centeredness is the predominant factor.

It is because of this reason that man, in his downfall as a spiritual being, became attracted to the animal form and became entrapped in it. Because man, in the spiritual sense, separated himself from his higher Self and became engrossed in the lower self or the self-sensations or selfishness, he attracted himself to, and was attracted by, that animalistic behavior which is predominant in many as of now, even to this date.

Not only have you been given this information and concept before, but you have been given many eons of experience in resolving them and evolving out of them. So, before returning to this information, I would prefer to go on into the higher elements and constructions that can aid in the spiritual evolvement and development of man as a spiritual being rather than as a beasted one.

OVERSHADOWING BY I AM PRESENCE

In the overshadowing principles of the higher Self or I Am consciousness, which is but one cell in the whole divine beingness of creation or Creative Force or the Father-Mother God, a certain elemental extract is to be understood. All has been revealed as to the properties of this I Am or electric body and the principles of its being of infinite relationship to Creative Energy or electrical substance, the fire of God.

In this procedure, when it comes in contact with the lower elements of Earth-made matter and substance you have a very definite reaction. It is fire upon the air which you are breathing and which is part of your existence. For without the air in your lungs you would not be able to stimulate the fire of being within, or the life force.

It electrifies the water element, which is the predominant part of your essence in a physical embodiment. It creates, through that electrifying force, a reenergizing of all the elements within the body simultaneously, as water is the surrounding or protective device through which all these cells are operating. It grounds you and it sustains you in this body. For should you become dehydrated, so to speak, your flesh form would turn completely to dust.

The earth or vegetable element or cells of your being become electrified and charged, and are among the first parts of your elemental body that are eliminated when the higher Self or spiritual being takes over in the physical, three-dimensional form or body. This allows for you to dematerialize and to materialize; and to walk upon the lesser elements without harm, such as over fire and water and through the gross matter of any kind of substance. It is the first demonstration of the total overshadowing of the light body or the electrical body, or the I Am presence within.

This fire or light, which is the I Am Self, can do all these things in any dimension and through any form it so chooses. It is via this element of fire or light that you will become transformed and will overcome every animal or vegetable or elemental property in the three-dimensional body. It only is through this I Am presence that you can sustain all life from planet to planet and dimension to dimension without a break in consciousness.

For if you pass into the soul body instead of into the I Am consciousness upon eliminating this physical form, you are bound unto the soul record or the soul evolvement, which is of a grosser substance than the light or the fire of the I Am presence. This light or fire always has been symbolized by the sun in our solar system.

MAN IS PART OF EVERYTHING

But it is more than this. It is the Son of God, which is the triune part of all creative fire and principle. Therefore, you are part of all that exists through this fire or element of electrical substance, which is energy, which is part of God.

When you have come to realize that you are part of God and part of all elements and part of all kingdoms, you truly are operating through the I Am consciousness or the spiritual aspect of your Self. You become enlightened or infused with the light of that being or substance which is love, light, life itself.

When you are living life, light and love in equal proportions through this manifestation of light or truth, you recognize that nothing is apart from you and that all parts are as one in the one and the higher element Which is the electrical force or Spirit Itself, in consciousness through you as an individual. I Am presence is but an individualization of one particle of that light, life and love. Therefore, you never are parted from life, light and love, but are one with it throughout all eternity.

INDIVIDUALIZATION OF CONSCIOUSNESS

As you become individualized in your own consciousness, and one with that consciousness that is individualized, you come to recognize all other parts that also are individualized, or parts of that eternal substance of life, light and love. In this consciousness you never are deceived, you never are separated, you never are alone in yourself, or without assistance. You never are complete unto yourself, for you only are part of the whole. The whole becomes individualized through a part of itself in you, your consciousness.

As you recognize this you recognize it in all others that likewise are individualized and imparted from the light, the love, the life force of God Himself. It is through this union that all truth can be made manifest, that all evolution takes place in time and in space.

It is not a single episode that concerns us or the I Am consciousness or the God Self, the God consciousness within the whole, because too much is involved, too many elements are involved, too many impartations or participations are involved in the conflict and in the growing and in the soul evolution. With this in mind you can appreciate why light and love and life are the triune principles through which all manifestation must become made individualized, why it must become solidified and unified.

UNIFICATION OF ALL

As one becomes aware of his individualization he becomes aware of his need for unification with all other parts, for he recognizes he is not complete within himself but is only a part of the whole. It is only through this I Am presence and consciousness that really higher, and the truest aspect of, unification or coordination and cooperation can take place in all matter and in all manners of development.

We have brought you a concept that needs much deep contemplation on the part of those who will seek separation and will not try for unification in the here and in the now and in the substance of the Earth body as it will have to come into light-body manifestation shortly.

You are being given glimpses of this light-body manifestation from time to time. You are being transformed by it from time to time. But you are not totally in it, for the time is not ripe, either for yourself as an individual or for the entire race and the program we all serve. So, until it becomes part of, and essential to, the entire meaning of what we serve—God, light, life and law—we shall have to go by it step by step and through it in agony, as has been man's way from eons of time.

But should you come into that electric body and see the unification and the oneness of all life, you can bring it about that much more satisfactorily and that much more quickly in a suitable arrangement and in an agreed-upon manner. For all are one in the light, and have that love and life within them that speak of this truth, and have this plan inbuilt in the I Am consciousness.

So it is this that must be taken into consideration at this point. All men who are in the I Am consciousness and who see as one, as sons of God, will see the One and the one plan and the one manifestation and will bring them about as one unified whole. There will not be many separate paths and many separate plans and many separate consciousnesses, but the one consciousness; which is the I Am consciousness within each.

I AM CONSCIOUSNESS

Since the I Am consciousness is the Christ within and since the Christ is without separation in itself and cannot be separated from itself, it can only be represented by its individualizations or its personalities of I Am consciousness within the whole. Your entire substance and growth as a soul and into a body form are to have this unique realization. For until you come into the I Am consciousness you are not aware of the separateness of your body personality, your soul function and your spiritual Self within. You think of them in three separate, distinct ways.

But it is by this coming into awareness of your I Am consciousness and the Self within, and this overshadowing substance of the electric body over the soul and the mind and the body of the personality which you represent, that you see the solidness of the one and the dissipation and the dissolution of that which is lesser or more gross in matter and evolvement, such as the soul, such as the mind, such as the body.

But it is true and it is recognized and it is known that when this solidification takes place there is that moment of fear—or rather, many moments of fear—and doubt when it seems as though a loss of something familiar is nearly the total experience or gain. Rather, it takes some moments and many, many experiments and the overshadowing of the Holy Spirit a hundred—nay, a million—more times in that substance and frequency to make you realize, and to make you secure in, the body form.

For if we wish not to have a body form, we would not form a body. But since we have formed the body and have preferred to work through the body substance and frequency of third dimensional matter, we expect the cooperation and the evolution of that form. We as the spiritual Selves, the light-body manifestation—or the higher Selves, if you prefer—have created these body forms and these soul records through eons of time.

FREEING THE ENTRAPPED

As given before, the Elder race or those who are in the fourth dimensional body, those light-body frequencies that dominated the Earth in the beginning of time upon this planet, helped to create the form and substance out of third dimensional matter. We, some of us, entered into those forms in order to experience, to experiment and to evolve out of them. Many then became entrapped, and so it has been an experience to free those souls and to realign them with their higher Selves and to re-fuse them with the light body from which they never actually have been separated totally, but have lost conscious contact with it.

That is the sum and substance of our entire program here upon the Earth in these latter days. It has been the sum and substance of our efforts as a race upon this planet and within this entire solar system for eons of time. The time has been given, as recorded by historical matter or evolutionary experience recorded as you now record the passing of events, as two hundred and six million years of evolutionary experience.

But we are in the latter days of this experiment and this experience. Many avatars and teachers, many prophets and many examples have been given in order to profess this truth and this knowledge and to bring about the end of this cycle of development and evolvement. Third dimensional matter, as it is now known on the Earth, is about to be transformed into the higher, etheric matter.

ALL EARTH MATTER EVOLVING

It is not only the property of man or the bodies of man as he experiences on this planet that will be transformed, but all matter that expresses in third dimensional frequency has to become evolved, has to become enlightened, has to become infused with its etheric substance. For all matter is part of an etheric, divine substance and energy. It in itself is descended into a lower or more gross form which you now recognize as the elements of the Earth. As these become refined and illuminated you will be evolved with them.

It is not so much that the elements become refined and then you benefit by them, but it is the reverse of this. It is that the consciousness of man, which is the controlling factor over the Earth planet itself, as sons of God, must refine the elemental substance or the gross matter and electrify it again, or reconnect it to its electric substance, in order to become refined and illuminated. By doing so through his own body, each man then relates to the substance or matter of the Earth in the elements that surround the Earth and which make up all other forms of the Earth frequency matter.

It is not enough just to give this information in just this way, for much must be experimented upon, much must be worked with in consciousness; and in some cases unconsciously, to the masses. That is why the I Am presence is ever working in and through all men—in the form of the Holy Spirit, if you will prefer that term —to inspire and to overshadow and to illuminate as many as can be inspired and can share in this information.

Some will do it in one manner or another. Some will go off on one tangent or another. Some will not respect all the teachings and information that come through another. But, at the same time, the I Am or Holy Spirit, the electrical substance of energy, will bring about certain transforming qualities within the grosser substance of mind, body and soul and thereby will infuse in those three lower aspects of being, that which is the highest and which is the transforming element upon the area known as Earth and its environs.

LIGHT WORKERS

You must begin to share in this with others of like substance, mind and companionship. As you will do this you will come to recognize that these are of a certain element of evolution in time and in space who can help bring about this transforming experience on the Earth planet, through the Earth planet and with all life form upon the Earth planet.

These are the light workers, the ones who will work with the light or energy or Spirit in themselves and in all areas concerning the Earth. These are the unifying principles or the coordinating light workers who will link together and will bring about an entire transformation of man, matter and life form as they express upon the Earth planet.

This is part of the mission which you have come to serve, and it will be executed as you have seen it. For you are in me and I am in you: Nada and Yolanda are as one. Nada overshadowing Yolanda will eliminate all that is less than perfection, all that is less than good, and therefore will allow only the light to radiate out from this center of being or manifestation which is body and form and to bring about that which is to be for the higher good of all; since all are in the One, and the One is in all.

So be it in truth. I speak in the name of Nada, which is a consciousness within the whole matter of oneness and which is an individualization of that oneness, God in manifestation. Amen. Om.

4. ENVIRONMENTAL RELATIONSHIPS

GUIDANCE FROM HIGHER PLANES

I am Lord Uriel, in charge of this [Yolanda's] body's functioning until the anchoring of light manifestation through and with this vehicle is completed. As in all cases, there is superior intelligence involved in seeking a proper anchorage and solidification through which you can be brought into proper alignment with the higher focus and function.

In the days of *now* which are upon you in the Earth planet and all planes and environs of that planet, you have a unique, special and unprecedented situation. It is entirely different than any other period in man's history upon the Earth. Therefore, we go slowly, enunciate carefully our desires, our will and our experienced guidance in order that you be not deceived and seek fulfillment other than through the proper procedures and channels of that higher guidance and final fulfillment for Earth and yourselves as individualized sons of God or lights in the one light Which is our Almighty Father-Mother Creator.

You have been prepared for eons of time and have succeeded in many avenues of expressing this very same function and delight of your higher capabilities. But because of the errors in the Earth and the manifestations upon the Earth of error thought forms and error projections and negative energies predominant everywhere, you cannot succeed in this mission without superior intelligences and guardians to aid you.

That is why we elucidate carefully our function and our plan and our place within your society and within your own evolvement and development as a race and as individualized beings. It is not, in any sense of the concept, to be derived in your thinking patterns as needing or lending or leaning upon the support, the guidance and the frequency of those who are in higher realms and in higher intelligent control of their energy frequencies than you are at this time.

For all are sons of God. All are capable of doing this same work and will be able to demonstrate and to teach this work in higher ways than ever before upon the Earth, when we have completed the initial stages of developing the consciousness and the body and of pressing out from the soul record that which has been of lesser events in the spiritual consciousness and concepts. It is this very process of pressing out your lesser energies and frequencies that involves the predominant controls, contracts, correlations which we bring to your minds at this time.

CONTROLLING THE ENVIRONMENT

You apparently are seeking higher demonstrations at all times. You are frustrated in innumerable occasions by the fact that your concepts do not measure up exactly to that which is able to be demonstrated through you at that specific time, in that specific place, regardless of what has been promised and what has been preconceived by you in prior lives, in this life and via the communications that are coming to you through these methods of channelship; not only this channel but many channels who likewise are indoctrinated into this higher manifestation and the preceding information that has come through about the New Age, the new bodies, the new heaven upon this new Earth of yours.

We wish to try to explain and to elucidate deeply what delicacies are involved in bringing about these manifestations, so you can cooperate with them and so you can coordinate also with the forces which are involved.

You are in control of your environment only insofar as you can comprehend the actual workings of that environment. You are part of that environment, regardless of whether you are aware of how to control it or are able to comprehend the deep intricacies that are involved. The environment in which you reside and participate is that which makes up your very being or natural force field of body, of mind and of soul. It depends quite specifically on your past experiences within that environment as to how much work you have to do in controlling your relationship to it.

Let us see that those who are in the environment of Earth planet and have been in the environment of Earth planet for many lifetimes have a better chance in some respects to control the elements of which they are part and with which they experience the same inflow and change of evolvement.

Those who come from distant places or those who have had infrequent demonstrations of bodily form in this particular environment, regarding the level of manifested frequency in matter—which is Earth plane frequency, naturally—have lesser chance of involving themselves with the frequency changes.

Because of this, many have come from distant places in an effort to teach the principles, but cannot be as effective in the actual demonstrating of these things because of this difference in experience and in relationship to the elements as they are produced in matter form upon the Earth planet at this time.

ELEMENTS AS PART OF MAN

Let us see specifically that the elements which make up your own body, your own mind and your own soul pattern are the intricate and specific items with which you have to concern yourselves at this very time. You are part of the elements and the elements are part of you. For this we deeply give thanks. This is the pattern, this is the manner in which Spirit has produced Itself in all things and has all things within It as one united whole concept and creation. Without this principle and basic premise we would not be able to transmute, to control, and to evolve in and through and with, the very matter in which we are karmically involved.

Therefore, those of you who have had soul evolvement many, many times upon the Earth have a unique relationship with the elements of this planet. You were created out of the elements upon this planet in the frequency form that now presents itself to you, or to your physical sensations; the five senses, of course. You as spirits were not created out of the elements, but the elements were created so you could have understanding and control and relationship with all parts of God, all manifestations of God or Spirit. From these very elements, gathered together in frequency form of one kind or another, Spirit infused Itself. This became the life form or the living matter of consciousness.

THOUGHT PATTERNS IN ELEMENTS

So, according to the level of consciousness that was created from the whole or the concept of God, you have the different levels or frequency forms or evolutionary materials with which to work and with which to relate. These elements all are part of mind matter, of Mind or Energy. These elements contain within them intelligence, responses and frequency vibratory modulations, all of this according to the level which would give it unique and specific development, evolution, growth, experience and the expression of God in myriad forms and manners for purposes of self-expression, self-propulsion.

As this is mind in matter, then it contains thought, for thought is a by-product or a process of mind itself. Nothing in spirit or in mind can be stagnant. Therefore, if all elements within any creation can contain and do contain thought patterns, you are confronted with this at all times in order to experience in it, to grow from it and to transmute through it. If this is the case, as it is, you now are confronted with thought patterns that have been impregnated into the elemental kingdoms from the very beginning of their incorporation into third frequency form or matter as now is called Earth plane functioning.

You have this large concept to contend with in the process of evolutionary growth from third dimensional matter into fourth dimensional frequency; which is the entire purpose of this program and plan we call the Mark Age period and program, or the latter days of man in Earth upon third dimensional frequency, going into fourth dimensional frequency, form.

This means that the Earth, all form upon it and especially mankind must evolve from this frequency pattern into a higher demonstration in order to evolve and to change and to control and to bring about higher expression of God, Spirit or mind matter. This again is because nothing must stagnate. All must be ever growing, changing and developing for the highest good of all with which it is involved.

MAN AND THOSE THOUGHT PATTERNS

Therefore man, who is part of the elements, and the elements which are part of the man form or the human form which the spirit of man has taken upon himself for expression upon this third dimensional plane or Earth frequency matter, have incorporated into his very being the many thoughts, patterns and experiences that have been expressed in, or which have been exposed to, the elemental kingdom of this planet.

Let us put it thusly. In the eons of development, as man—who is the Son of God and the co-creator with the divine principles of the law of God—came into this frequency, saw this creation and participated in it, he has presented it with many thoughts, both of good and of evil. He has experienced many trials, many tribulations, many civilizations which rose and fell and which were re-created in another form or in another place according to the higher justice and the plan of those superior intelligences which guide and help man to come into his highest expression as sons of God or co-creators with divine law.

Through all these eons of time, the elements—being of lesser intelligence than man, who can transmute and change via his superiority over all other forms in the planet—came to accept, or to incorporate or to have imbedded within them, these thought patterns, these thought projections, these thought experiences. Some of them are imprinted so deeply upon the elemental forces that it will take eons of time to unfathom the depths of them and to re-create or to reassess what is proper and to supplant it with much higher, better goals and concepts.

But in the meantime man is involved in the planet. He is deeply engrossed in the expression of the elements as they take form in one area or another, as they express animal and vegetable and mineral life in one form or another, and even has it within his own house or body concept. You eat of the minerals, the vegetables and the animal kingdoms. Therefore, they are totally incorporated into your cellular structure, and the vibratory rate of the thought patterns which are inculcated in these various cellular structures are part of, and developed within, your entire experience.

MASTERY OF SELF

It is often given unto you in metaphor that man is his own kingdom—his body, his mind and his soul—of which he is the lord and the master. This is absolute truth. But within this concept you all will have to come to the conclusion, and to the ability, of controlling every single aspect that is part of that kingdom of which you are master. You are to be master over every cell, over every thought, over every organ, over every part of your being in mind, in soul and in body.

But because you have within the kingdom of yourself many thoughts, many experiences and many avenues of expression already negating the Sonship, negating the mastery, negating the desire for good, you are torn asunder in your body, in your mind and in your soul; because you have participated in, and have ingested all of, this material through eons of time.

It is for this reason that certain souls incarnating in certain sections of the planet where there has been especial evil or error or negation or downtrodden experiences cannot rise easily above them.

For the very air they breathe is inculcated with the very thought patterns that have existed there from time immemorial, as you may express it upon the Earth at this time.

CHANGE BY THOUGHT

But all can be changed again through thought. For thought is the main material with which mind works, and mind is spirit. You are, first and primary, spirit and only are participating in third dimensional form or matter as an experience and as an experiment in time and space. But you exist above and beyond that.

Since you do exist as a spirit, and your higher Self and spiritual body always are above and separated from this unclean condition, you can work through the light body to transmute and to change and to cleanse not only your own individualized body or cellular structure or self-expression but all that you participate in, including your environment, your fellowmen, your elemental kingdoms and eventually the planet itself.

It is of this that Jesus spoke when he said: ye shall do even greater things than I have done. It is because he understood the principles involved and was able to control many situations in his own environment, in his own body, in his own fellowmen and in the elements around his particular area of power, that he could say this and give you courage that collectively all of you together, conceiving of this one truth and divine principle that thought is matter of mind, could be given this encouragement and prophecy. For together ye shall do much greater things than he as a single lighted or Christed soul was able to accomplish in his individualized mission during that particular incarnation as Jesus of Nazareth.

It is because we of the angelic realms, particularly this aspect or area of control—Seventh Ray function over devic forces and kingdoms—can bring to you greater strength and application. It is through the Seventh Ray projections of divine love and peace that we can express a higher dimension in the turmoil of life as it expresses in third dimensional frequency matter. In all the other areas of soul and spiritual growth there are activity and change and movement. But in the Seventh Ray activity over the devic kingdom it proceeds from out of the stillness and the oneness and the uniqueness of unity of principle or Spirit matter, Mind truth.

Give this then some thought as to how you can begin to infuse your own soul bodies and physical functions with this light, through Seventh Ray function of love, light, peace, truth, rest. It is through this function that many can begin to unify themselves with one another and seek a unified plan on the Earth to bring about these great changes that are needed in the present era or time of *now*.

TRANSMUTATION OF THOUGHT PATTERNS

You have been given many principles by which to work, and many more will be given to you as time unfolds for this presentation and program matter. Not all is to be accomplished within the next twenty or thirty years of scheduled time. But much must be conceived of and projected into the Earth planet in order to change the orbit or the definite cycle of cause and effect that has been in operation for eons of time. As we have indicated already in these immediate goals and lessons herein, there is a procedure involving elemental kingdoms and the evolvement of these elemental kingdoms from the past. The date has been given to you as two hundred and six million years prior to this time.

Therefore, it is necessary for you to conceive, and to accept eventually, the fact that in that two hundred and six million years the life course or cycle of cause and effect has been in operation according to certain thought patterns and evolutionary developments. These are going to be extremely difficult to change or to transmute. For once the wheel is put into motion according to a set frequency or pattern, it is not changed unless mind matter changes it, unless thought projections dissolve it, unless determination and understanding resolve it.

So, the wheel is turning. Until man in his present experience as a son of God, but living in the elements of this turning wheel of cause and effect, is willing to change the frequency motion of that wheel of cause and effect, he will not experience any change, nor will any change be possible. For it is man who has created the turning of the wheel of this type of frequency matter or cause and effect, and it must be man which changes it. Such is divine law in action.

Therefore, it has taken all these eons of time for man to begin to realize his control over the elements, his control over his Earth planet, his control over himself. For he is of God, made of the sum and substance of all attributes of God; though he is not God Himself, but part of that substance. He must act accordingly and demonstrate according to those principles involved in spiritual life. Since man is spirit, then he must perform as Spirit performs: divinely, justly, lovingly, willingly and in unification.

This unifying principle is the magnetic force which you call God or Spirit. It is through this magnetic force that all things are cohesively drawn together in the oneness of God or Spirit or matter in manifestation. Therefore, through this you become one, in this you are expressing according to the frequency of your own individualized Christ Selves and Christ bodies, which we call the light body or the light frequency. Light is that electric force or that electrifying light or fire that brings about an individualized soul.

CLEANSING OF SOUL PATTERNS

In the soul pattern you have many changing conditions also. In the soul pattern, you have incorporated within, or ingested within, many of your past episodes of experience, not only on this planet but other planets. You have incorporated within you all the thought forms you have experienced anywhere and everywhere through an eternity of gradual growth and evolvement. You have all this to call upon. You have all this to unfold.

It may not be necessary to unfold each inch of the bolt of your own consciousness or soul expression. But that area which is to be exposed to your conscious mind to overcome and to transmute will be brought to your conscious attention during the time of evolution or change of frequency. For as you go into the light-body manifestation and as you call upon the soul to cooperate with that light-body frequency and manifestation, you are exposed to all those areas, experiences, relationships and thought forms in which you had participated at one time or another.

Therefore, please be aware of that which comes to you in these days and through your inner Self expressions and thought patterns. Be ever cognizant of these thought patterns, soul functionings and soul revelations. Where there is error, cleanse it through new thought form, new thought patterns, new mind manifestation. It is yours to do. It is the glorious pattern and the glorious control and the glorious deed God has bequeathed unto His sons and daughters of the Spirit, which you are.

In this aspect every soul must become cleansed before the light body can manifest fully in and through its projection of self or body or personality as you now express. Until all these things are presented to your conscious mind for evaluation, for possible change and for uplifting you cannot sustain the light-body form for a very long time upon the Earth planet or in the frequency or form of third dimensional matter.

It is only by this very process that you can come into a fully manifested light-body frequency and a sustained anchorage of the light body through the physical form or the soul pattern that is also expressing, or is the expression of, the personality self or body form as you now know it.

It is not so much that we are concerned with the body, with the personality or with the soul. But since they are the lesser frequencies and are the accumulations of all lesser frequencies during your eons of experience and are solidified or crystallized in a certain pattern of growth and understanding and have within them all the elements of those things which man has experienced or has projected out into thought, and these thoughts are becoming things or are expressing in cycles of cause and effect, you have to confront them and to be one with them and to work through them and to change them from what they were and what you have participated in as a son of God.

IMPACT OF THOUGHT PATTERNS

In summing up, it is important for you to understand that because you are a part of the race of man and because you participate in every element and in every thought projection sent out by the race of man, everything that exists is a part of you to one degree or another, or as much as you will let it be a part of you in one degree or another.

Therefore, the thought, the action, the word of any man any time, anywhere in spiritual creation has had some impact upon your entire racial picture or inflow of life. It is because thought is never dissolved unless a higher thought consumes it that this can exist in exactly this manner. This is for good, and this is for evil as well. You have before you the great masters, teachers, avatars, demonstrators and the higher intelligences everywhere in higher planes and planets giving you the benefit of their experiences, their thoughts, their words and their deeds projected out into the ethers or into the elemental kingdoms for you to pick up.

The very air, the very water, the very Earth products you participate in are impregnated with all these higher goals, teachings and manifestations. Likewise, the lesser ones or the evil ones are part of your participation. So, it is a matter of changing or redirecting, or consuming by the light or fire of your higher intentions, that you dissolve the error that is impregnated everywhere in and through and around the planet by all these higher frequencies of which we are speaking.

MAN ON EARTH MUST CHANGE IT

You will be given much to aid you, and have been given much to aid you, in the days to come. For we are entering into that period when the third dimensional matter definitely will be transposed or will be transformed by the higher thoughts, deeds and goals of mankind present upon the Earth planet. The goal, the deeds, the thoughts of those coming from higher planes, projecting through Earth channels such as this one and the many thousands all over the world who are aware of and accepting these principles and ideas of the New Age or Mark Age period and program, can help in this process.

But the higher deeds and goals and thoughts of those not incarnated upon the Earth cannot do the work. They only can inspire the ones who are channels upon the Earth to broadcast and actually to effect the changes that are to come. So, it is your responsibility again to work with those who are so inspired and dedicated and to help bring about that which must come, regardless of how much is to be sacrificed of the lesser or lower evolvements and demonstrations. Many things will suffer in this process. But this is the growth pattern and this is the precluded process that have to be experienced and determined.

Go now in peace, in love and in truth. For I am with thee, all of thee, in every aspect that is conceivable. It is because of the vastness of this program, but also of the single-mindedness with which we of the archangel and angelic forces can determine, that the work we have to do is simple. For there is no equivocation or no lack of dedication on our parts.

It is only in the transforming to your level that the problems can be engrossed. So, be not of this type of manifestation, but be of our level, our superior intelligence and guidance, where we are one, where we are participating in the single control and single guardianship of all that is, all that ever was and all that ever shall be: God, our Father, in us. Amen. So be it in truth. I am Uriel.

5. FALL OF MAN

FROM THE CENTRAL SUN

I am Nada speaking. I am sent from out of the central sun to speak unto thee who are upon the Earth planet and all who reside in astral dimensions around this solar system. You all have been in the center of creation wherein your souls were formed and your bodies made manifest, wherever you experience life and form. You all may recall this coming forth, in days hence, as it is part of your unfolding and the rending of the seventh veil which is to be rent throughout this solar system before the cycle of this time of *now* completes itself.

Since it is my strictest intention to bring forth that which is recalled and spoken of in many ancient scriptures and prophecies, so that each thing may be fulfilled unto the dotting of every *i* and the crossing of every *t*, let it henceforth be given unto you in consciousness and spoken of clearly in these days of *now* and in the time right before us.

We set forth to be Ancient of Days, the Jehovah consciousness of the one God Which is our Father-Mother Creator. In this aspect of unification with the All that is Spirit, that is complete and that is all-knowing, we broke forth into experience and expression, pressing our lives into existence wherever it pleased us as children or creations of this one Spirit. In It we were never separated; not from It, Which is in divine intelligence, thought and mind and the power of all existence, nor were we separated one from the other.

Though we knew it not, we were capable of separating ourselves from certain aspects of consciousness. It is not reality, this so-called separation. It is a mirage or hallucination, that we are separated. Like unto any dream we experience in our mortal existence, so is the separation from God. It can last eons of time.

The separation began when this solar system was created. All of us, who were sons of God in the spiritual sense of it, came unto this solar system along with those of the devic and elemental kingdoms to produce a haven for all life-form experimentation. We drew upon our experiences in and through the fire or the element of being which is the etheric consciousness.

Our setting forth from the central sun then was naturally into the sun of this solar system. Therefore, all who are part of this particular solar system and the solar systems relating to the central sun of the many solar systems which surround the central sun have come from the same source or level of creation or beingness. It is many eons or ages gone by, long before historic memory of the sons of God or the Elder race could recall and have recalled.

However, it is impressed upon every element and every iota of soul expression in and through all these beings and in and through all these forms; and is especially impressed in the etheric substance, which shall come about and which shall become our whole and total indication and identification soon, through the evolutionary process through which we are going at this moment in space and time.

EXPERIENCING OF FORM

We came into the sun of this solar system and began experiencing form. Some resided in one form or another upon the various planets of this solar system, depending upon their own individualized exercises.

Out of the Elohim, which create substance and matter and elements, we each have our relationship and regard and responsibility. It means, therefore, that we are related to one of the seven aspects of God consciousness or Godly functioning, as to which part of the seven parts of God form or life substance we took, or from which we were taken. This is the individualization of the sign or the substance of life. This light then becomes love or form, which is the Sonship of God.

When we recognize and realize our source and substance we are forced to experience and to express it unto every iota of our existence. It is the form of pressing out that which is from within, or the origin of being. This substance takes on form depending upon the elements that are able to be created out of the areas or dimensions of cycles, frequencies and levels of being.

The Elohim press out toward the archangel or celestial kingdoms to control and to solidify this consciousness or intelligence. It is through the elemental kingdoms, under the supervision of the archangels, that certain elements are able to exist, depending upon the area in which it is controlled and contemplated.

Therefore, you have dimensions, you have varied planes of existence within those dimensions. Out of those differences of planes and planets, life form can become existent. The life form takes place through the substance or matter of those frequency elements in those dimensions which create various planes of existence and produce a planetary frequency upon which a physical, or formulated physical, life source can become treated.

The sons of God or the individualized consciousness of the Elohim then try to supervise or to supplement life programing not only for their own experimentation and learning process but because, whenever there is an area of expression of creation or creative substance in form, mankind must have some experience within it. This does not mean mankind inevitably must fall into the trap of its form and existence, but must have supervision or exercise over it.

EXPERIMENTING WITH LIFE FORM

In all dimensions there are varied frequencies that can produce certain forms and through which certain intelligences can express or treat themselves to additional programs of life and light and law. Through this the sons of God have learned how to experiment with life form and matter, regardless of what a dimension is or the planes of life which are created out of the varied dimensions or frequency substances.

All creations are not alike, nor do they have the same source of, or length of, life. Some are very short in range and in purpose. Others are very long ranged and subtle in their purposes. Therefore, it becomes necessary for mankind, which is the Son of God or the Christ light, to have discernment and purpose in controlling, producing and evolving through it with that particular source or life form.

What happened in this solar system is that many souls who took on these experiments and experiences became very enthralled with the variations of life or length of life through which these varied frequency matters could express themselves and experience various sensations. Not all were three dimensional, for that came as the last and final, lowest step as far as mankind or the sons of God were concerned.

It was upon this planet Earth that the last and final step was taken for man to experience the lowest form or lowest rate of vibration for a physical form as far as he was concerned; a Sonship, that is; not as far as life form can exist, for life form can exist on lower frequencies than this and with less potential than this.

But in our particular solar system, as the energies became lesser and lesser and the varied life forms became more conclusively physical, the sons of God became more enthralled with this area of development and so produced a slower rate of evolvement and a slower evolutionary pattern than had been produced anywhere else in the galaxy or universes that revolve about the central sun from which we all originated.

When this became a necessary factor it then was necessary and proper to bring about a form and a life substance known to us all as the Adamic race. This was part fourth dimensional and part third dimensional. For out of the third dimensional frequency—or the so-called clay or earth matter—of this particular dimension and planet, a form was acceptable into which the sons of God could enter in order to experience evolution.

MAN NOT FROM LOWER KINGDOMS

At first, when this planet was created in all of its glory and beauty with a series of elements quite unlike any other dimension within this solar system, the various sons or lights were attracted to the elements and entered into the elements and experienced life form through them. This is not the same as saying man evolved up out of the elements. But because man did experience through these elements and participated in these elements, he has a race memory of being inside, or part of or participating in, the very elements of which his physical form is made up. It is out of this that he must now evolve.

In quite specific terms, let me be precise; as I recall, and can give evidence of, this. Many upon the planet are having recall of the same and have not understood, or rather have confused, what they have experienced or have remembered. One can enter into the water element and become part of that water element and experience very many things that the intelligences which make up the water elements do experience.

One also can participate in the earth elements or mineral elements and say and feel that he or she is part of that mineral element. In other words, many have thought of themselves as rocks or part of an underground cavern or the very minerals that spring up and can nourish or discourage further life form as you now know it in the physical, three-dimensional body.

Many entered into the later experiments, or animal forms in the levels of their creation from the fish through the bird and through the crawling and walking creatures of this planet. But that does not mean the souls or the sons of God, the light or Elohim, participated in an incarnation through these things. Because mankind is an etheric or spiritual life and light he can participate in any kind of form or matter for a length of time, depending on his ability to sustain intelligence and control over that. Many can do it for an hour or two, and some even have been able to do it for weeks at a time.

Through these various times and experiences we have a great many legends, or what have been called fairy tales, out of the imaginations or memory patterns of certain souls who were sent to create this imaging for mankind in order for him to understand that out of the etheric substance of his light body he can and must participate in all elements of this planet and the elements of all other planets within this solar system before he evolves back to his Sonship or the Son light in this solar system, then reaches up and out of this solar system into the light of the central sun.

FINAL FALL

This is a long way off. This consciousness or concept will be taught later on in the New or Golden Era of our time, when the Son of God or the light of this solar system is able to shine through each one's consciousness through his Christed elevation or his etheric body. This will be taught by the Elder race in returning back into the etheric substance and in the memory recall of all mankind upon the Earth planet. For many will come from all over the solar system to teach and to recall this matter for each one who is here entrapped and has been so entrapped for millions of years.

It was out of this varied evolvement that we had the original separation when the Elder race tried to control and to teach the lower manifestations, or those sons who had fallen from this consciousness and experiment all over this particular solar system. All then became collected for a time upon this planet or schoolroom which we call the Earth plane. This means the astral as well as the physical plane. For at that time there were constant recall and regression from one plane into the other and progression out of one plane into the other, depending on how much progress each soul could make.

The Elder race then became entrapped by the conflicts and difficulties of raising those delinquent souls from out of this. So we had the fall of man, as we now know and understand it. This final fall occurred twenty-six million years ago. At that time the very lowest elements of man's consciousness, which was almost bestial in contact and context, were left to his own resources. Various plans and programs were instituted to raise him, but not within the same dimension or upon the same frequency form as we had tried to resolve in the previous cycle of time; from two hundred and six million years prior to this date or time cycle, to be more precise.

When this occurred we had a varied type of civilization and plane. From this came many schisms in the thinking of the Elder race and in the consciousness of man as he evolved upon this planet. Some did graduate or did resolve their particular differences and came unto the various planets or planes which revolved around those planes and planets of higher frequency and development.

But not all were able to come into the highest or etheric substance which was theirs to have in the original state of grace, or from their highest form or God Self consciousness. This is the reason, in the total history of this solar system, we have not been able to raise the Earth and the Earth plane dimensions much higher than they have been, and thus to raise the entire solar system out of its third dimensional frequency matter before this time.

MEMORY RECALL FOR ALL

But since the cycles upon cycles are ending and we are ending not only the cycle around our sun, but the cycle of two hundred and six million years around our central sun, we now must see this in its proper proportion. For this reason those of the Elder race or those of the Son light, etheric substance must redescend into the Earth plane and bring about a sudden and very unique and special conscious effort to reawaken, and to lift the veils which have enshrouded, man's memory and recall of all these things.

This is the very reason you are having so much difficulty upon the Earth planet at this time. For it is the memory recall and the error conditions that are coming to face each soul from eons of time. From the beginning of his error conditions he has created these thoughts or forces. They remain in the substance and matter of the elements all around the planet. It is very unique for man to understand this in a three-dimensional consciousness, or in a material kind of consciousness; a so-called logical, intellectual consciousness which he has prided himself upon for such a long period of time.

This is very shortsighted, for the intellectual aspects of man are only regarding his one incarnation at a time, whereas his soul recall is that which has been going on for eons of time in his development or reevolvement out of the matter consciousness in which he now finds himself. Regardless of whether he remembers this or accepts this, it does exist. It does affect his progress. It does impress upon him as an individual and makes for his efficiency as a soul and as a personality.

But whether he accepts it or not is not the point. It exists. It affects him. He must come into conscious grappling with it at one time or another in the evolving out of material and intellectualism into the spiritual and etheric substance which he is eternally and which shall be his final stepping stone into that which is his etheric body or his Sonship with God.

CAUSE AND EFFECT

His thoughts, his patterns and his experiences are impressed wherever there have been life, light and truth. This means that wherever man as a son of God has produced an effect, he must reap the consequences of those effects. He must reestablish or recall all that has been set forth from out of his mind or matter. Wherever he has thought, he has to rethink or to remember that which he has thought.

It is out of this pattern that he has created certain forms of civilizations, and that certain elements or forms of elements are interacting with him as an evolving being or race. When man has experienced violence in a certain place, or has brought or has caused violence in a certain place, the very elements are impressed with these thoughts. They become the form and substance out of which his creation or society is developed. It reacts upon man, regardless of how long it must take, for man is the control over all forces and substance.

Whether he likes it or not, this is truth. Whether he accepts it or not, this is the law. Whether he wants or not, he is brought back to that scene or society, or a similar one, in which he has participated. It concerns him not only upon this Earth but throughout this solar system. Therefore, if a man or a society of men has created good, he or they will reap that good in a similar situation somewhere within this solar system, whether it be on this plane or planet or on another plane or planet elsewhere in this solar system.

It has been so according to divine law and cannot be excused in anyone's case. For this is the process of evolution and experience and the executing faculties that man has and under which he must learn to function. When man has produced these error thoughts they have a reaction upon the very elements in the area where he has produced them. Thus you have good soil, bad soil. You have good forces in the mineral kingdom and bad forces in the mineral kingdom, which help you or hinder you in your own particular way of experiencing.

If you have produced a cold front of thinking, you will live in a cold front or area of climate similar to that which you have produced. It is only out of this that man can really relearn that which he has created out of thought and has made into matter. For all things are just in the law of God and in the ways and workings of that law.

Thus there are some who are born unto advantageous states of consciousness and planes of existence and have many good things come unto them out of the natural resources of the life and the substance of the plane or the planet where they reside. Others have quite the negative. But because we are in the ending of an age and time, we must come face to face with all these elements and thoughts and creations and substances of these thoughts, and work with them and re-create them.

CLEANSING OF ERROR

Therefore, it is very necessary upon this planet for all those who have created error thinking—and perhaps, in some ways, good and productive thought patterns—to come into conscious grappling with them at this ending of an age and learn to dispense with them and to re-form and to recalculate them for a higher good and a higher evolutionary experience. Thus it is that the planet comes to its final and lasting judgment. It is this judgment—not that God has placed upon man, but which man has placed upon himself through his eons and eons of experience—where we now find ourselves.

Where you have sent out good, there you will reap it. Where you have sent out evil, there you must find it, change it, transmute it, climb above it and dismiss it once and for all. The planet must be cleansed regardless of how many must suffer the consequences. For the time is ending. The period of grace is over. This is the last judgment for all who have created life and life substance here upon this Earth planet.

Therefore, many are coming under conditions which they find abhorrent. Many are finding great and wonderful opportunities upon this planet. For there are great and wonderful new scientific developments which have been earned from eons of time, and from varied experiences on other planes and planets also. Therefore, some are having great and wonderful episodes and challenges and opportunities to grow and to evolve and to understand, whereas others are suffering untold scars upon their soul memories and consciousnesses. These are rebelling and these must be aided. For only by projecting out into the mass consciousness that these are the end days and these are the final opportunities can you hope to salvage as many souls for the coming New Age as is possible.

Many shall not be reached because they do not want to be reached, or they cannot be reached because of the accumulation of their errors to such a degree that it is impossible in this ending of a period of time to save them from that. There must be then a place for these remnants, and a place shall be secured. Many shall have to be conditioned to accept that place and plane and planet.

But it will not be within this solar system. For there are great reasons for this to become an example unto other solar systems and areas within the universes controlled by the central sun of which we speak, that are beyond the ken of this type of comprehension within the soul memory of the race of which we now are speaking.

LEMURIA AND ATLANTIS

Many are going through the stages of evolution they should have gone through in the days of Lemuria and Atlantis. But in those days there was a great deal of segregation of the souls. A type of caste system was put upon the race at that time, especially in Lemuria twenty-six million years ago unto the two hundred and six thousand year cycle when Atlantis was formed.

Those that were in the Lemurian period were able to see that there were souls who experienced the various elements and enjoyed the experiences of those elements and wished to continue to enjoy the experiences of those limited elements. Where we had souls who had partial control over their etheric substance and partial enjoyment of the three-dimensional form that was being experienced and experimented with upon this planet, we had many, many forms which were able to be created. Many types of animal forms, many types of elemental forms were being utilized in order to arrive at a compatible or equalized, harmonious energy frequency or dimension.

In the process of devolution you have this peculiar series of partial etheric substance and partial physical substance. This was the so-called Garden of Eden in that period of time. Some souls were able to be in both at the same time and to experience both at the same time. Because the Elder race—or those who remained in the etheric substance, for this is what the Elder race symbolizes—did not wish to participate in the third dimensional form, they created what is now known as a form of caste system.

This caste system remained in the consciousness or in the creative memory pattern or soul consciousness of man as a race. When the planet broke up into various continents—as it had to, due to this separation of ideas—many still were enforcing a type of caste system. This is the Eastern cult or culture that we have even today in the society or the civilization of man upon the Earth planet in third dimensional form.

In the Western society or evolvement, or the development from out of the lowering into third dimensional form, it became more equalized. So you have what you call your Western concept, or the attempt to equalize all under the same form, law or regulations.

Thus you have the two types of concepts, the two varying ways the Elder race tried to cope with this most difficult problem. As man fell more and more into, or became entrapped into, the third dimensional frequency form—which was by the time Lemuria was over and had to be destroyed, for it had not worked out in this caste or gradation system of partial frequency in varied grades of expression —we had to discard that concept totally. So you had the breakup of many types of societies. Today you have remnants of those societies still existing in various climates and countries of our planet which we call the Earth.

When the experiments of Atlantis were permitted and man again rose into a form that was more compatible or equalized than he had experienced in the Lemurian civilizations, we tried, through the equality system or equal opportunity system of Atlantis, to bring about the resolution of third dimensional frequency form rising up into fourth dimensional frequency matter, which is the etheric substance of man's eternal being. Too many wished to dominate in this particular civilization or series of civilizations which lasted from two hundred and six thousand years ago until twenty-six thousand years ago.

At the sinking of Atlantis, twenty-six thousand years ago until twelve thousand years ago, you had the attempt to bring about a reorder and reorganization in the present form that you now recognize or know him as a race. So begins the history of the present civilization or societies which now are prevalent upon the Earth. You have many variations upon the planet in very many climes and projects where societies, civilizations and countries are in conflict with one another. But you do not have a varied form of man as you did in the prior civilizations, particularly in the Lemurian period of life upon this planet.

REEVOLVING

Now we come unto the present, and we must see only from the present. The past history of man upon this planet is really of little consequence in these latter days, for he will not live in this consciousness but will pass through it. Much as the embryo of each soul in a physical form must pass through the various stages which he has experimented with in the evolving of this planet, so must the embryonic form of spiritual matter and consciousness pass through these various thoughts, patterns and expressions in the reevaluating and reevolving into science of mind and spiritual light form.

Man is moving from a total three-dimensional form upon the Earth into his final stage of fourth dimensional frequency form upon the Earth planet. As he experiences this evolving process he will remember and will recall some of the past, even if only in a dream or in a momentary flash of recall or an identification through meditation with one form or another. This can be very fleeting.

In a moment of reflection the most stolid, sincere and materialistic type of person can identify with a flower, a drop of water, a cloud, the air or an animal form. That is sufficient unto that soul for that time. It is not that he evolved from out of these forms, but that he will identify with his experimentation in this form and formula of third dimensional form since the days of his beginning upon a third dimensional frequency form.

When he does this it is sufficient unto his evolving. When he comes into a higher experience of light and love and life he at least has reached his etheric substance or his spiritual consciousness for that moment and for a matter of time. But it is not sufficient, for him to come into that for a fleeting moment. He must continue to evolve into that consciousness and in that respect for long periods of time.

Whether this be done in his sleep and reflective state is simply up to him, how fast he is willing to grow and to go into God consciousness. It is not a matter for anyone to teach another, or to demand of another, how long and how much he is to do this. For each must be given the opportunity to know this exists. This must be part of his reevolvement back into his true substance and form so he can bring himself back into true consciousness of his spiritual Self and his higher, evolving state of being.

It must be reemphasized here again and again that man has not evolved out of the elements, but that man helped to create or to bring about certain elements in each plane and planet of this solar system so he could experience and control each one of the elements and each combination of those elements throughout this plane and planet. He has done this by entering into, through his etheric substance, those elements and those combinations of elements so he would know how to exercise control over them and be the executor of his Father's estate; which is Creative Energy, in the substance and form and matter of it.

But since man has not remembered himself as a son of God and has not regained his control of etheric life or spiritual consciousness through a third dimensional vehicle form, he must do it in a very slow and evolutionary process; as he must go through each step in reverse of that which he experienced, or brought upon himself, in the devolving process of falling into matter and creating for himself a form that was suitable for his soul and spiritual substance or etheric light form.

COLLECTIVE SOCIETY NECESSARY

Man is coming unto his true light form. Man is reevolving from third dimensional frequency form into a light substance upon a three-dimensional planet. But when he does create a fourth frequency form sufficiently satisfactory and sufficiently controlled and anchored in sufficient number, he then will transmute the entire planet by his collective consciousness, his collective society, his collective desire to do this which was his original premise and proclamation from the divine Godhead or the Elohim from which he first originated.

He may not try to produce this result before he is in his etheric consciousness. He may not accomplish this until he is collected by number in sufficient quantity to bring this about as a collective society. He may not do this until his Prince or leader is returned unto him in the form of Sananda, who last incarnated upon the Earth as Jesus of Nazareth.

This type of history is well within the contemplation of all who are serious about the evolution of man upon this planet for the last two hundred and six million years. It is not an arbitrary figure, it is not an arbitrary proclamation or claim. It is mostly within the divine fiat and law of man as he is producing and projecting from etheric substance into physical form upon this planet at this time.

It is within his responsibilities to know and to understand that something very unique and particular has occurred upon this planet, throughout this solar system, and from all solar systems within the confines of the central sun from which we all became a united being, a united and collective society of sons of God.

HIERARCHIES

Within the creation of God is a series of hierarchal forms or set-ups. None are exempt from this law or divine method of operation. All are part of, or participate in, this kind of systematic evolving consciousness and responsibility. When one is aware of the different phases of this he usually is quite cooperative and willing to participate in whatever level of the hierarchal development of creation he has come to serve. For within the consciousness of each one is the realization that all are one in the One and all are equal within the One, although each is a part of the One and none is the whole. Each is but a part of a whole, and the whole is the sum total of all its parts.

This is why the two concepts are equal, and served equally for a time. The caste system worked for a time, when one realized that he participated in various areas of evolvement, development and service. Others were willing to work within that concept as long as there was no superiority of one form or caste or organization over another. But where it became a matter of, and where it still becomes a matter of, superiority or individualized control over another form or part of the whole, then we have error and confusion, and a great deal of relearning to do.

It is in the conscious application of the principle that all are equal within the whole, for the whole is not complete without all of its parts participating, that we do have what is a more satisfactory form of society, consciousness and responsibility. That does not apply only upon the Earth or in the societies or governmental responsibilities but is the consciousness within the evolving development of men as sons of God and as the real co-creators with God of all His kingdoms throughout all eternity.

For man as a race or as the Son of God does have the opportunity, uniquely so, of experiencing all parts of creation in every dimension or form the Creative Substance has made possible throughout eons and stretches of evolvement, throughout all kinds of creations, here and elsewhere.

There are suns upon suns from which man can have experience and through which man can experiment, but not until he learns the evolvement through hierarchal control and responsibility and not before he is willing to see all parts of that Hierarchy as equal substances or equal opportunities to serve.

It does not mean, either, that man will experience every single part of hierarchal substance or form. It does not mean that man must enter into every combination of form and substance throughout all kinds of creations and areas of contemplation. But it does mean that he can become aware of them according to his own individualized choices. For man is a free soul as long as he realizes that his freedom is to express Spirit, Which he is.

TRANSFORMATION OF EARTH

As a son of God, or the creation through which God experiences Himself or Itself, man can become a child truly of the One Who is all, and can inherit any part and all parts of His kingdom according to his choice and his desire for experience. If man desires the kind of experience he has had upon this Earth, and wishes even less in opportunities that are upon this Earth, he will get exactly what he desires. For his desire is thought and energy and form. He then will create a form, a substance, a place for himself. This is the law of cause and effect. He will bring it about for himself.

But there are some who wish to rise above this that they have experienced for so many eons of time. Collectively, there are enough to bring it about. For by the strength of their collective desires this program shall be accomplished. In accomplishing this evolutionary process they will become as sons of God and will live in the light body upon this third dimensional frequency form. Then, in the collective form of their fourth dimensional frequency or etheric substance, they will transform this entire planet back into its original state of spiritualized energy, with form that is pure and true and light and love.

Let it thus be our collective and unique and united desire to see this come about as quickly as possible, with as little pain as possible and with as many as possible that are willing to do so. Although it will take a minimum of one hundred and forty-four thousand souls upon the Earth, in this unique consciousness and desire, it does not have to be limited to that number. For the sake of man and his soul evolvement, let it be all who will desire embodiment and will be given embodiment and can rise above embodiment into spiritualized frequency or light substance, the I Am consciousness, the I Am which speaks through me as Nada of the Sun. Amen.

So be it in truth, for this is truth. I have spoken it thus for all men on Earth. Unto all men I address myself. For all men I thus dedicate this as a beginning and as an end result. For by speaking the cause and by bringing this information into conscious condition it becomes a fact accomplished upon three-dimensional frequency form, and is being transmuted by the very effect of these words upon all in all dimensions that concern this Earth planet. Amen. Amen. Amen. So it is done on all levels of consciousness. I am Nada.

6. SONSHIP WITH GOD

ALL IS OF SPIRIT SUBSTANCE

I am Lord Uriel. Speak thus in my name and bring light unto all which is in manifestation upon the Earth planet and its environs. You have been told of many wonderful things that are transpiring at the present time. It is becoming and necessary that you learn how all these marvelous and myriad miracles of life substance and sustenance can continue for ever and ever.

You are created out of life, light and love. You are the essence of Spirit, and Spirit is the only essence of which you are created. In these many manifestations that appear to your eyes and senses there are varied components and formulas that derive themselves from out of the spiritual essence of which you are but a part. But all parts are the same in that they also are consisting of this life, light and love substance which is Spirit Itself. However, in the complexities of expressing this spiritual form on a three-dimensional force or in other aspects of life, living and expression, you have a combination of multiple parts and functions that do not seem the same or alike.

In many other instances you have the same condition resolving itself in an area of life and light where there is no form that is perceptible to your physical senses and sensations. Yet, they exist and have expression and evolve and become a part of your subtle environment and being and nature, regardless of this fact concerning your physical perceptions. But you are not aware of them, nor give credence to them, nor pay attention to them on the higher intelligent level which is your conscious concept of what is taking place. However, it is mine to assure you that, regardless of this, you are deeply and essentially touched by them.

This is how we come to inform you and to evolve you from out of this mesmerism of third dimensional living at the present time. We want so much for you to be able to appreciate all that exists and to get substantial assistance from all that exists. But before you do you must come to recognize that all that exists is essentially and predominantly and foremost made up of Spirit substance and sustenance. This clue informs you of your real eternal life and gives you encouragement and procedures by which you too can appreciate that eternal life; and exist in it and evolve through and with it in an intelligent manner, such as you are endeavoring to do at this particular time.

Since all matter of Spirit is created by thought, you reach this essence through your own thinking and feeling natures. Therefore, it is most essential that you return unto the essence within yourself which is Spirit in order to commune with, and to appreciate, that which is of Spirit in all other forms of life and creation.

MANIFESTATIONS OF GOD'S WORD

In the beginning was the word, and the word was with God. God spoke the word, and so it became manifest. How is this to be considered? In the beginning was the word. But the word had to be formed out of thought. For a word has within it the implication that words are outer symbolizations or outer manifestations of an idea. God is that idea, or Creative Substance is that idea.

The word, then, is the single individualized idea taking form and root via a spoken decree or thought manifestation. This is how it all began and this is how all life form took on form and existence. Even the sons of God were but thoughts or ideas within the one idea or mind of God. From these individualized ideas, or subsequent ideas within the one idea, came the word, or the formation or the speaking forth or the bringing forth of those ideas.

Each child of God is an idea within the idea of a Sonship with God. Therefore, when God spoke and made manifest His one single idea of mankind, which is the Sonship or all manifested form, through the Mother-Father Creator or the equalized balancing poles of existence, He made a body of life or light or love which is the Sonship or kingdoms through which man may express. Out of those kingdoms man became a Son.

Those sons within the Son became the levels of the one single idea, or the many, many varied forms or functions within that one idea or form. Each man has his own single function or formula to perform. Each man or son within the Sonship of God is unique and individualized and personalized within this one idea. So, although you may conceive of, and go back and forth into, the Sonship creation, of which the idea is mastery and Godlike production, you still must perform that which is part of the entire idea of Son or manifestation of the light and of the love and the life form which you are.

In the beginning was spoken the word. That word burst into myriad forms or functions within that one word and idea. For man must go forth into all creation and kingdoms of the Father-Mother God and experience all parts of that creation and bring it back into its one whole, united, cohesive force, and work within that united whole. This is the function and the being of the Sonship; and, as a matter of fact, all levels of the Sonship or all creations within the Sonship.

CONCERN IS WITH PRESENT

So, man has his very unique place within the scheme of things. His responsibilities are very great. But beyond this we cannot give words or ideas or orders. Let them be your own to achieve and to understand and to receive from within, because only in that way will you reach that Sonship. Only within this Sonship can you achieve that which you originally were created to perform. That is to do the Father's will in the Father's way and in His kingdoms throughout all eternity.

For as Jesus of Nazareth had said: in my Father's house are many mansions, and I go to prepare a place for you. It is this concept he achieved and comprehended that made it possible for him to encourage those of his day to keep seeking and working, regardless of not understanding all they were achieving and all that would be achieved in time immemorial before them. For it was sufficient unto the day, the work that beset them of that time; not to be so concerned with future goals and works and kingdoms that were beyond their immediate attention.

So it is true today in the minds and in the works of so many who wish to have a great deal more of the information than is available to them; and yet they cannot conceive of, or work with or achieve, that which is before them as of the moment. They fall by the wayside and fail in their missions in that they cannot do this work that is with them to do as of the present, but seek always higher and different and more complex terminology and works and concepts.

So, it is here a warning that, though we are giving much of the past and the evolutionary progress of man upon this planet and in this solar system for eons of time, and yet are opening the vista of your thinking to many more creations and kingdoms and ideas than you possibly can conceive of or work toward in this lifetime and in the many lifetimes to follow, we still are not asking that you seek this that is so far beyond you that you neglect that which is right there before you and in such dire need of work and conscious application. Give this your most serious consideration, for it is the most important part of what is being given herein and delivered unto this channel for your edification and elucidation this date and heretofore.

You have been in the creation of the solar system long, long before it was able to take on a solidified form. You were spoken of, and created out of an idea, long before form filled the void of all life and light and truth. So, your existence goes beyond all that is ever able to be conveyed to you in word or in form. You will go on unto eternity experiencing, experimenting and seeking ever higher truths and learnings in your evolving grace as sons of the living God.

RESPONSIBILITY TO LOWER KINGDOMS

Out of this living God you are life and experience life, and give life unto others as well. You are the life-giving substance to those who are of a lesser kingdom. Not that you of yourselves create these lesser kingdoms, but by giving them a purpose in their creation, by being part of their purpose in creation, you are to them as gods; as we, in many cases, seem to you to be gods before the throne of God.

Since we have achieved our oneness and understand our unique function and can educate and guide you, we seem to you far beyond that which you can achieve. Therefore, you seem the same to those who are of a lesser creation, and seek always to please and to express themselves in your own domain. These kingdoms I speak of here are the animal, the vegetable and the mineral kingdoms over which you have dominion and over which you have caused much disgrace and disorder and chaos. Therefore, you must learn to be as their guides in a proper way and to be their way showers in a very high and spiritual manner. For, of them Spirit also is the essence and is the whole and the total sense of triune creation.

Out of this you must become more spiritualized yourselves. In seeking this unification and this overseership with them, you will gain your own higher evolution and truth and principle. For from them came the being and the need to sustain a life form through which you yourselves would have a form and an existence in this particular dimension. They remain in a certain frequency orbit only to sustain a life and a substance, a plan and a planet and a place of abode for those of you who are the sons of God. Yet you heed them not and conceive not of what you have done in the way of neglect and disregard and disturbance.

Out of all this they retain the thoughts you have had for eons of time about them as creations and about your own creative life. They are like the soul substance or record keeper of your own experiences. From out of all these substances and elements you reap that which you sowed into their consciousness, and are experiencing the very playback of your own words or thoughts.

For the words you spoke were the thoughts you as co-creators were able to manifest or to project out into the essence of life. Just as the Father thought or had the idea of a Son and then spoke the word for the existence of all matter and form, so the sons of God as co-creators and imitators of this Creative Substance thought the thoughts, spoke the words and helped to create the forms which now exist everywhere in their own kingdoms and abodes.

INTERPLANE COMMUNION A NECESSITY

Let it be your primary objective to come into communion with that which you yourself created, which you have helped to put into motion upon this plane and upon this planet. It is up to you to get to know those forms and those creations and to come into conscious contact with them through the thought projections and the thought communions and the inner substance relays which are the Spirit essence of you within and the Spirit essence of those creations within them. That way you become one and reenact that which you had taken upon yourselves as co-creators in God's domain.

As you do this you can cleanse and re-form and reshape all that has taken place in this particular solar system, and particularly on this planet where you now reside. For though you have much power and acknowledgement within the entire solar system, especially those of you who have had evolvements and experiences and roles and missions upon other higher planes and planets in this solar system, your main responsibility for the present embodiment and time is with the Earth. Otherwise, why would you have been created in a form or an embodiment for the present?

It is the Earth with which we are mostly concerned. For as we can cleanse and raise the Earth planet and all life form upon it, so we can raise and form a new solar system from out of this same development and goal.

Let us please be appreciative of those who are working on higher dimensions, and in other planes and planets as well, for this same goal and purpose. As they come to you in thought and in projection of thought and as they come to you in some form of materialization via space beam and spacecraft—which is a legitimate form of communication and projection—then you can come into unification and communion with them, thus raising yourselves and aiding them in the work they do in cooperation with you and all life form upon the planet.

This is all that is required of those who are upon the planet at this time. For your range of difficulty is limited for the present time. Your range of effectiveness is certainly going to be limited until you overcome the difficulties of your present area of dispensation and development. You have been given this indication on numerous occasions and thus must begin to put it into practice and to evolve from out of your difficulties and to achieve these goals where you are.

Similarly would you have a student, who is subjected to limitation in a certain area or course of learning, staying or concentrating upon that before graduating him into a higher school of more complex learnings and more complex schedules, so he does not become unnecessarily confused and thus upset the equilibrium of those with whom he is associating. This is pretty much the same principle that is involved. You will appreciate it, for it is common sense, and one which you can easily apply to your present status and the quotation of your majority members of workers and light projectors.

LIGHT-BODY MANIFESTATIONS

In all this work there are the goals of higher substance and matter ahead which you shall evolve into and shall be able to perform. These are the light-body functions and manifestations. Since it is not a learning process by which you can rise up and grow into all these applications via the systematic process of development in a gradual unfoldment of this Christed Self, you do not need that type of systematic or grade-level instruction.

Your Christ Self or light body always has been and always shall be, regardless of how you face it and regardless of how you relate to it in the physical embodiment. It is the light body or Christed Self which originally created the form in which you are presently embodied.

Therefore, you do not evolve out of the physical into the light body, but you grow accustomed to the physical and have the light body overshadow that physical at the time you are able to fulfill all that is required of you in the physical embodiment; or by reaching the consciousness through the physical to that which always has been and always will be, the light-body manifestation, you come into the full realignment of the light body overshadowing or consuming that which is the physical body and personality, or the mental broadcasting unit station of the personal self, in truth upon a set plane or promotion.

You are required, by this promotion, to live in varied dimensions, planes and planets. It is by projecting, from out of the light body, a suitable vehicle through which you can experience these varied places and myriad functions that you become used to these varied vessels or vehicles through which you do operate at different times of your evolvement and development, and thus are promoted from one level of function and one level of responsibility unto another.

It is not that you earn them from the level where you are, but that you conquer the level through which you have been projected from out of the light body or Christed Self manifestation. This has been the unfortunate mistake that has been propounded for eons of time upon the Earth planet; that mistake being that you evolve from a sliver of self into a fully, more all-embracing conscious aspect of Spirit, which is the light body or spiritual Self. It is the other way around. The all in you always was, always has been and always shall be. It is. You are. It cannot be otherwise.

LIGHT-BODY CONSCIOUSNESS

Therefore, as you enjoy the expression from this light-body consciousness in the present environment where you are, you earn further responsibilities and conscious applications of that light body in the dimension or on the plane where you are existing, and thus change that plane and that dimension. In other words, the sons of God from the beginning have had the responsibility to experience every possible form of creation which has come out of the infinite variety of combinations that are possible for a Christ or for a light body to overshadow.

You are in this stage at the present time. You are in this particular transition of one stage into another at this particular time. In other words, you are at the ending of an age, when third dimensional form as it relates to this particular planet within this particular solar system has come to its fullest and final stages of expression. Therefore, as you are, and have been for eons of time, experimenting with, and experiencing life and control over, this three-dimensional physical form and planet and all life form upon this planet, you now may come into your own and true and original state of form, which is the light body or fourth dimensional frequency.

The fourth dimensional frequency form, as we express it at this time, is the Sonship or the light or the transcending essence or substance which is the God Self within that is, always has been and always shall be yours and your unique participation in all creation throughout all eternity.

As you come into conscious awareness of this in multiple varieties or numbers of consciousnesses and individualizations, a majority or a significant number of such consciousnesses can bring about the change that is required for this plane, this planet and all life form which comes under the supervision of those who are the sons of God on this plane and on this planet. This is your total desire; and has been your total desire from the beginning, when you first took form upon this planet or saw this planet as a place or a schoolroom for which you could have experience and energy expression.

TRANSMUTE AND TRANSCEND

Let us begin the long series of evidences that shall show how you may bring this about and overshadow all that already is existing on a certain frequency modulation, which is the third dimensional frequency level and the physical planet as you know it. It will not be destroyed in the sense of being eliminated, but it will be consumed in the sense that it will be overshadowed by, and will be absorbed into, the higher Self function. So, it is not that you lose something, but that you transmute and transcend from out of the use of that level and that expression.

There are many who are not ready for this. Therefore, they will sense the loss or will fear the loss, because of the lack of understanding. But because you are able to demonstrate it and to teach it, those of you who are advanced enough to read and to appreciate this information and can relate it to all other history of man and scriptural references, we seek the aid and the assistance of this unification of those who are known as the light workers, or the co-operators with the Elder brothers who are here in every level of possible assistance to give this the final and lasting uplifting steps necessary to complete that which once was started and since has been evolving from out of the Earth and into higher avenues of experience.

7. FREE WILL

UNIQUE FUNCTION OF LIGHT BODY

The service and the function of the light body are very unique, in that it serves to unify all substance and matter throughout all creation. It is that which is related to the spiritual essence of all created matter, whether on this dimension of perception or on other and higher, or even lower, dimensions beyond the senses of perception where you are presently residing. This unique service is that which can bring about a change in consciousness and a change in form, for all source is one and all energy force is the same.

However, the degree by which that energy force is utilized makes for the difference in levels of consciousness, functions, services and responsibilities. That is why the sons of God or the light workers on the Earth are particularly responsible and needed at this time. For they can comprehend this and can utilize it a lot better than any other forms of creation who are residing in physical embodiment upon the planet with which we are concerned.

You therefore are implored to be about our Father's business in coming into as much cleansing and rectification of the errors that have been perpetrated for eons of time, under the guise of spiritual workers and unifiers of God's kingdom. For since the beginning of life upon this planet man has inherent within his spiritual pattern or consciousness, as a race embodied therein, that he has this responsibility, this goal and this function. How he has served is the trying question, and the crying warning for eons past.

FREE WILL IN EVOLUTION

Throughout all scriptural references and spiritual intunements man has been placed at this unique head of his solar system. It began when he came out from the central sun and explored all the universes and galaxies that are a part of that creation. His varied functions have included setting forth a hierarchal government where it would be able to control and to constitute a systematic procedure of growth pattern, growth planning and programing of the highest nature, or of his spiritual nature.

Therein man had his own free will to choose and to make these patterns and to project these programs, of which we have given some indication through these informational discourses; but not sufficiently to plug up the present time and the present level of program efficiency so that man can utilize his still active free will and consciousness in cooperating and coordinating with a higher and more exclusive detail and schedule.

In doing this he is challenged as to whether he can reach a higher intunement from within and seek his Self-enlightenment and his Self-revelation of what is proceeding upon the planet and throughout the solar system at this time. For without this it would be dogmatic and dictatorial. None can be satisfied with this, for that is not growth or progress in man's evolutionary state in the third dimensional form.

We had this problem in the earliest days of those who descended in the fourth dimensional body as the Elder race, or utilizing only their light-body consciousness, overtaking those forms and parts of the race that were to experience only third dimensional frequency form and to be given plans, programs and proper procedures to evolve out of it. Naturally, they had been the effect of many causes throughout the many systems; and were not placed just arbitrarily in that position because of some higher design to bring about this experience, but because they as souls had brought about this condition within themselves.

The Elder race, or those in the fourth dimensional frequency body, seeing this need upon this planet for an evolutionary program, set about to bring it into a manifested glory unto the one Source and the substance or essence of all things. However, the argument that ensued and the two diverse methods of bringing this into effect were so severe that some of those even in the Elder race fell into matter form, or into the third dimensional frequency, in order to experience it and to show their domination over it and their displeasure with their fellowmen or brothers who had a slightly different approach to this.

That is cause and effect, and free will, in operation. Never can the principles of divine law be overrated or sidestepped in any manner whatever. Therefore, these souls are still in the evolutionary progress of determining which source of right and wrong, which source of programing they prefer; exercising thereby their own free will of choice and of being independent within the race consciousness and being independent from God consciousness. This always is the privilege and prerogative.

HIERARCHAL BOARD AND FREE WILL

But by expressing these thoughts and giving this history we program you to understand further and more intimately that a Hierarchal Board condition does not preclude a divine fiat or system whereby only this method or way or truth can truly exist. It is given with the understanding that among those serenely placed in these positions and given these responsibilities and functions there resides the course of freewill choice and employment.

It does not mean, either, that some are evil and some are good, for this is totally in error concept. When God has bequeathed the freewill principle within His kingdoms to choose their way of experience and expression, it is part of the divine heritage within of choosing and of expressing and of evolving, as the law is just and the law is true and the law is inevitable. But by these same tokens —truth, justice and righteousness—all are going to evolve back into the state of grace and love which is the divinity within each one.

So, regardless of whether a program takes two hundred and six million years or twenty-six years, it is not of concern to the spirit within. For the spirit within always is divinely balanced and just and filled with love and truth. Therefore, it is in existence in that state. Only that part which is the free will or soul of mankind can deviate from that truth, that balance and that love and beauty.

We have tried to give this example in order that you have homage and respect for the Hierarchal Board of our solar system, which in patience and in grandeur has decreed many of these set policies throughout our solar system with which we now are concerned in the evolutionary pattern and progress.

But it is not beholden to the Hierarchal Board or the spiritual government of this solar system, which makes these decrees and sends forth these program plans and matters, to unfold them unto each one or to force each one to adhere to the exactitude and the principalities inherent within those laws and fiats of spiritual consciousness. For the only thing that is inevitable is the oneness of God and the eventual balancing and truth reavealment within each creation of itself, of its source and of the essence with which it has been created. Nothing can ever change this, since it is divine.

We give you now these examples so that, when working with divine law and heritage, you give as much of yourself to the divine within as possible. For the freedom of will is that which can will over to the higher Self all that the higher Self wills to do for itself, or the God Self within. It is extremely subtle, is it not, that the free will really is expressing in the will of freeing itself to do God's will, or being in the God-given consciousness of oneness, truth, balance and proper unification?

NO FREE WILL IN LOWER KINGDOMS

It is not so in the lower kingdoms. They do not have this consciousness, to enact a defiance against the inherent pattern of their evolution. But they are deeply swayed by man's thought pattern, since man is the protector of, and the domineering influence over, their growth and progress. So, in a sense, when man holds back his own progress he holds back the progress of all the lower kingdoms as well. For they are influenced by his thoughts, by his desires; and, if it is so, his defiances of those laws and progressive reports.

But now we must come into repeating that though man is related to the elements via these lower kingdoms, these lower kingdoms are not going to force man to become one with them. They cannot have this control over man. In other words, they do not have the will, nor the intelligence nor the intuitiveness, to be able to project themselves out, but must wait as receptive vehicles for that which comes to them from man.

It is having this Self-knowledge, Self-appreciation and Self-contemplation of his own divinity and of his own creation that marks man as a special species, upon the planet and throughout all creation, from what is not of that level or of that hierarchal development.

THE SEVEN ATTRIBUTES OF GOD

Here is the reason the hierarchal concept is the divine one. In the setting forth of Its own manifestations, the Spirit Which is God has made all things according to a hierarchal or a progressive, graduating concept and evolution. Therefore, one type of creation is merely one set or combination of functions and services, having within it that much of God plan and God concepts but not all of the varied concepts and proclivities of a God-given attribute.

These attributes of God are seven in number. They are called the divine principles. They are expressed in a series of seven tributes to the essence or nature of God Himself, or Spirit, as It is in existence; as It has been and as It ever shall be, without change and untouchable. [See chart on page 139.]

These seven come forth in seven ideas or words or manifestations called the Elohim. They are as follows: one, the law or the will of God; two, the mind or intelligence, the thoughts, of God; three, the feeling or love nature in God; four, the manifestation or development of manifested form of God; five, the integrating and unifying principles that are cohesive in God; six, the transmutability or change that is everlasting and ever so in the forms which God can bring about; and seven, the resting or peace aspect or the love of God Itself; God's love for His work, and His work for God Himself, or Spirit in action.

These seven principles or attributes are ever in motion. In every single creation it follows that certain levels are given one or more of these attributes or principles. Wherein the archangels are set forth from the Godhead to protect all these manifestations which come out of the word or idea of the seven principles or Elohim which are God, they do create a series or forms which make up the elements, or which we call the devas, of all the universes.

DEVIC KINGDOMS

These devas are not individual consciousnesses, nor do they have functions as personalities, but merely are expressing in elemental forms regardless of the plane, planet or dimension where they can take root or can bring forth fruit. These devas are not in any way conscious of the work, but are solely responsible to the angelic kingdoms for that which is performed through them.

Then it is the sons of God or the Christhood—which is the third principle or the third aspect of God; the love nature or feeling aspect of the intelligence and the law in action together, creating then the Sonship—which take on or mold from the devic kingdoms those elements that can bring forth material substance for whatever plane, planet or dimension that Sonship wishes to have manifestation. This is the hierarchy of God's creative principles and never can be changed anywhere or any time.

According to the need for expression or experience in controlling these seven levels or attributes of God principle, the different devic kingdoms are called into play to produce some form or substance out of which man then molds a life or a society. This goes far beyond anything that can be brought to conscious or mortal attention at this time. For the men on Earth are so far removed from the memory pattern and the development of this original concept of their beginning as a race upon the planet Earth that they hardly can conceive of what is inherent in these words.

However, the Master Jesus prepared you, in these words: if a man have faith and he say unto that mountain, be ye removed, that mountain must be moved; have the faith that this can be done. He told you not how to do it. He told you not that it was your inheritance and that you already had created that mountain from out of a higher aspect of understanding and faith and knowledge. But he prepared you for the coming days when he would return unto the Earth and teach you how to bring about these principles.

AGE FOR REVELATIONS

It is now those ending days when all that has been hidden from man's conscious memory is to be revealed. All the veils that man has created for himself in falling into the trap of materialism and conscious application of physical life, we can proceed one by one to rip away; the veils of ignorance and doubt and loss of faith in himself as a son of God.

Though I give not strict rules and regulations at this time, for it is not mine to do, I prepare you as he, the master of this plane and planet, has been preparing you for thousands and thousands of years. Though you have not the entire historic record of his responsibility to this planet and to the men who are entrapped in this plane and planet, we can give you some hint of what is before you, what has been behind you and what can come after this time of trial and tribulation.

But if you will have faith and you will go within to the essence or the source of what has been given here, that place within where you are one and all memory recall is the same as since the beginning of time, you will find here the words and the concepts that strike the inner chord and make you reveal unto yourself the truth of this so there is no dogmatism and there is no force in use when it comes to accepting, or to believing in, these principles.

Be thou trustworthy of this confidence. Be thou responsible for these thoughts and revelations so you can begin to receive from the divine source within your own Christed Self and life body that which is yours to bring into manifestation in these latter days, or the time period we call the Mark Age plan and program.

APPLICATION OF DIVINE ATTRIBUTES

You have been given these things by many sources and enlightenments. But now is the time to put them into practice. It means that, by recognizing the source and responsibility of the oneness of all Creative Principle in you and in all life force around you, you utilize this principle and unification to bring about a more equitable justice and a more cleansing process for the inevitable day of oneness and truth and transmutation.

You go from one step to another in this process. As you apply each of the principles or attributes of God to your own life, you see them in the ensuing relationships with all other sources of life. Then you bring about that justice or balance between you and all other forms of life upon the planet. This brings you into that responsibility of consciousness which you as sons of God will have and always have had, and must have again in this day and in this age.

Then all together we can work as one, for we all are part of the one Source. In these latter days ye shall do greater things than he who was Jesus did in his time. For ye will do it in concerted action, in cooperative action and in the coordinated principles and plans that are revealed as Hierarchal Board decisions.

For the Hierarchal Board of our solar system, which is the spiritual government that has brought forth these divine ideas, has done so not out of arbitrary thinkings but by the collective meditative processes of what is the highest, best and most conclusive plan for the greatest number of souls entrapped in this solar system during this period of time since the coming out into expression within this solar system.

Most other planes and planets which have evolved to an equitable and enjoyable relationship with all life form and substances, in which they equally are enjoying the creation of God's expression in them, have achieved such in recent times, according to historical events. We mean this very specifically to say that in the last two thousand to ten thousand years have all the strands been placed in proper position.

END OF GRACE PERIOD

But the Earth has not. In a sense, the Earth has been given two thousand years of grace period to do this. With the coming of Jesus to the Earth planet in that particular form and personality—though he had been on the Earth many other times giving similar principles and ideas—he had allowed a grace period of one more age of Earth time; or the equinox revolution of that planet around the sun.

Now that this has come to pass and the time has been completed as far as solar system developments are concerned, there is no more grace period left for man to achieve this consciousness of his own free will. So, those who are responsible in the angelic kingdoms and in the spiritual consciousness as enlightened souls or the Elder race, who have not fallen into the trap we express as the third dimensional mortal existence, bring a new plan, a higher plan that is to effect the goal in as immediate a time and in as short a time as we possibly can achieve it without upsetting the equilibrium within a certain number.

THE ONE HUNDRED AND FORTY-FOUR THOUSAND

The number of souls who must maintain their equilibrium in incarnation is the one hundred and forty-four thousand enlightened. Of these hundred and forty-four thousand enlightened, many are of the original Elder race, not entrapped therefore in the third dimensional frequency consciousness.

This means that although they have the body and form of the third dimensional frequency consciousness, they had not become entrapped in this dimension. Their soul records are cleansed or unblocked from any past procedures or complications that would have prevented them at this time from enacting the roles they shall be called upon to enforce as part of, and combinations of, the seven attributes of God in action upon the plane and planet with which they are concerned at this time.

So, you will have many who proceed in a manner that is quite uncommon to the usual on the planet in these latter days. You shall recognize them not, for they proceed as ordinary individuals. But out of the wake of their enforced light and truth and single-mindedness of their missions and roles you will see a wave of great change and spiritual upliftment that has not occurred on this planet for eons of time.

Before you commit these to another era and to another nomenclature, be sure you understand what has befallen, or has taken place upon, the plane on which you reside. For they have come from distant places at great sacrifices to do this work, and at the bidding of the Hierarchal Board of this solar system to do their special jobs and to enact their very especial formulas that have been eons in perfection.

Go and give this as much thought and contemplation as befits a true child of the one source and substance Which is God. Give this every consideration until it becomes a conviction within. When that occurs you are part of the plan, part of the programing, which has been sent forth from out of the mind and the heart and the love of God.

I am Uriel, now befriending and solidifying the light bodies of those who came to do this work and to bring about this unifying principle of God in action upon the planet Earth in these, the latter, days called the Mark Age period and program. Amen. So be it in truth; for this is the essence of truth, and the law that has proceeded from out of the word which is I Am. Om.

8. DEVOLUTION

MASTERY OVER ELEMENTS

As we have unfolded the progress of mankind in his spiritual evolution, we have used both the words *evolution* and *devolution*. It is connoted, therefore, that one is going forward and one is going backward. Yet, in all serious contemplation, is it not to be accepted that all is going forward if good or marvelous lessons can be learned in the progress?

So, let us now examine step by step that which is the devolving into matter. But in this evolutionary story of mankind we can see how much good will be derived when the mastery over all these elements finally has been resolved and has been put into good practice and for the highest manifestation of the true Sonship with God which is man's ultimate and fruitful destiny.

When he entered into this solar system in the etheric body, which is the divine essence or substance of his spirituality, he had the obligation to see all chemicals and matter-form as a part of his playground or schoolroom for learning. Through mind matter and control he had the ability to juggle these essences or to formulate them for his highest ongoing and experience. He had the ability to taste of them or to utilize them in his own experience, as we already have evidenced by the fact that any spiritual essence or any part of God can enter into any other part of God and can experience that particularity of the God substance or essence which is Spirit.

Thus man did participate, in his etheric or light substance, in all the elements that were abounding throughout the solar system. In their varied component parts or combinations he had myriad experiences with them. As the angelic form saw fit to help sustain different frequencies or forms of matter for evolutionary progress also, he participated in this governorship over those forms and matters. So, he became one with all things that exist. In the solar system he was able to entertain every dimensional and vibratory frequency that exists. Believe us, there are more than can be counted by your understanding or comprehension.

Still, they remain; for man is still part of the governing body of this entire solar system. He has the ability, in the etheric substance or light-body manifestation, to go from many dimensions into any one he so desires, with the process of thought control and proper thought procedures of his mind-over-matter abilities. This he has done, and this he still may do, in any dimension or in any frequency that pleases him at this time.

SUBCONSCIOUS TRACK

You have had such experience in meditation, in contemplation, in dream, in memory recall and in the many or myriad types of legends, prophecies and the like throughout historical record on the Earth planet and in the memory subconscious of mankind. This has confused so-called material-minded individuals, for they cannot respond to, or consciously comprehend, what it is that is being dragged into the conscious mind from the subconscious recollections.

This is the major point of difference we wish to clarify for the time being; that is, the subconscious track and the development of the astral form from out of the subconscious track of mankind as an individual and as a race throughout all eternity. As man in his etheric substance had taken upon himself the experience and exploitation of many of these substances and forms or chemical divisions throughout the solar system, he would have a series of experiences that naturally were recorded into substance or mind matter.

There is a level of mind or a function of mind or memory which is called the storehouse or the subconscious track. This storehouse is a very real and alive pattern or part of man's entire being and nature. It is called the submemory part; or, the psychological term, subconscious; or, in the spiritual connotation, the astral form.

The many and involved conscious experiences that no longer can be controlled, or conceived of, in conscious application at the moment are put away, or stored temporarily, in the subconscious track or storehouse. This has created a form or a substance which becomes part of man. It is of the lower frequency and is of a lower response within him. But it also is of a more tenacious nature than his material aspect or experience.

In other words, the momentary experience is the material substance. The momentary sensation or realization or lessons become the conscious level of comprehension or development. As you dwell more and more on that aspect of it, and less and less on the purposes behind your reasons for experience or in storing these experiences for future exploration, you dwell more and more on the conscious or physical level.

But as for the subconscious function or astral function, you must have a place or a process by which you can recall and relive and relearn all that has been given. This is the astral form. So, man in his divine essence or spiritual being, the light body which is derived from out of the fire elements of the God consciousness of being, had produced this astral element, which is of the air elements of matter.

Into the air is stored all these experiences of the individual, of the race and of the collective consciousnesses who work together. This aspect is true not only of man but of any form that exists, of any area of consciousness that collects itself for an experience or for exploration of God substance or light-fire itself.

Therefore, this air feeds the fire. The fire is not extinguished by the air, but the fire or light of God exists eternally and supremely over and above all such things and never can be diminished by any element or thought consciousness or devolution of the matter or the substance of the God light-form.

ASTRAL AREAS

Involving himself by this process time and again, man created the astral areas of experience. Many shun, and shudder at, this concept. Well they should, for in this storehouse of memory and experiences are many undesirable experiences and events. But it is only from the God consciousness or the light comprehension that man can reach into, examine and change that which is occupying that area or that substance of matter.

You must do this, each one individually and each group, as a whole. In some cases, such as on the Earth planet, the astral area has become so infused with error events and with conflicting and confusing reports of what has taken place in the evolution and history of man in this solar system that it becomes the predominant part of his entire threefold being: the physical, conscious; the astral, subconscious; and the etheric, superconscious or I Am Self beingness of God.

In the Earth level it has become well known that the astral forces or energies are those which are predominantly clouding the inflow of higher light frequency and form, and are dominant over the physical, conscious aspect. Many on Earth would like to declaim this consideration, saying that man is in control of his life, his experiences and his explorations on the Earth fully in the conscious;

and that man has taken upon himself this very conscious deliberation of his life and of his society and of his future; and has not looked into the past, nor would he look into the past.

But unfortunately this is totally erroneous. For he is subconsciously controlled and he is overshadowed by the astral events and the astral consciousnesses of those who have been done wrong or who have been deceived or who have been in conquering positions throughout eons of time upon the Earth and who refuse to relinquish these areas of control.

Therefore, your higher spiritual teachers and the informants from the superconscious level desire to ignore or to step away from any contact with the astral associations. This in itself is self-deceiving, for you cannot eliminate a condition just by ignoring the condition. You must grapple with the condition. This is the condition that exists for the latter days, or the Mark Age period and program.

But let us not dwell too long on that aspect of it, since you will have other instructions which can guide you into this matter to handle them. In each one's life he is grappling with them sufficiently well, because his own life stream or energy from within, which is the God substance, is forcing him to face his own past relationships, consciousness, ideas and factors which make for the confusion and upsets in the planet as you see it today.

Also, this same condition exists in the elements or in the other kingdoms of Earth for the present time, since there is great upheaval, and there will be even greater upheaval in the near future, among the various kingdoms of Earth which are under the command of mankind at this time. For they are infused also with this effluvia or record storehouse which has been recorded collectively by the mind and the material events of man, and all the events and experiences and evolutionary processes within their own species throughout the history of time upon this planet.

The same applies to all other planets and places of experience and expression throughout the solar system, as it exists throughout all creation. But the reason the other planets have not become so infused in material matter as the Earth is that they have grappled with, or have comprehended, their astral forms in a much higher and more equalized manner than any on Earth have been able to do.

DEVOLUTION AS EXAMPLE

So, in the devolving of man into physical form upon the Earth planet and in his process of evolving up out of it, mankind as a whole is learning a tremendous lesson, which is the correlation and the corresponding force of astral upon physical, overcome and overshadowed totally by the spiritual or I Am essence of his divine creation; which is all that ever is, all that ever has been and all that ever shall be, in truth. All these other areas or lesser forms are part of that, but not the whole of it; nor is it clouded.

In the process of man on Earth learning the corresponding relationships and control aspects between Earth, astral and the higher force, he has set an example and gives for the rest of the race throughout all time and eternity an opportunity to see how far down the scale of experience one can go and how soon it can be worked out of his evolutionary grace.

This is why in the opening statements of this particular chapter I have explained to you that even in the devolution of a race such as man, we are evolving the entire concept and the entire form of man throughout all this solar system; and, as a matter of fact, throughout all creation. So, there is a service that is being performed and a function that is quite necessary for the experiences of the sons of God which are inherent in the sons of man on Earth.

Again you can realize and recapture that this episode we now bring to your immediate and conscious attention is an evolutionary pattern and is a necessary lesson for all. As man learns every nuance and every part of his eternal power as a Godlike creature or a son of God, he then can teach and example it unto all others everywhere. So, much good is going to be accomplished from this.

THE DESCENT

Now we come to the area or the era when, in the process of developing this storehouse during his eons of expression and experimentation, he saw upon the planet—and upon the other planets within this solar system, incidentally—that the physical form was quite desirable or interesting to experiment with. This became the momentary pleasure, or the sensations which man as a race—or some of the men within the Elder race—could and would descend into for a temporary period of time.

It did not start as an entire incarnation. It did not start as a desired part of his evolutionary or learning process. But because he is the master over all these forms and the governor which controls all the interrelationships and the processes of developing new and varied creations, he expected himself to have a part in them; and a direct part, at that.

So, many within the Elder race who were of this inclination then descended into the matter form that was being presented to them for study and for control. This is the parable of the sons of God entering into the daughters of Earth and becoming part of them and cohabitating with them. This took many, many eons of time for it to become a solidified or an extemporaneous jurisdiction.

In the legends which are passed down in the Greek and Roman literature which you call myth and legend, you have many such episodes, where the so-called gods and goddesses were able at times to enjoy the fleshly pleasures or the material aspects of the planet. But they would ascend again into their higher realms of expression and not continue, nor in the beginning could they continue, this life.

They thus were able to enter into any form which was being created out of the essences and the varied levels of creative substance for their experimentation and learning process. Thus many who were brought from all over existence, because of their varied falls, had taken on these forms in one way or another and were then part of this planet's existence as a schoolroom.

THE ENTRAPMENT

It was here that the Elder race, or that segment of the Elder race, finally made its fatal movement or fall in that, instead of gradually teaching and governing them to throw off the trying control of material matter and subconscious or astral form, they entered into and enjoyed some of the sensations of this plane and planet.

This did not take place to this degree upon the other planets of this solar system, where the distance or difference of attitude remained intact. The teaching and learning process of control over the material substance and of desiring the cleansing through the astral form or the memory block has become, and still remains, a true and clear line of evolutionary progress in these other dimensions or frequencies, which also are part of the same school or evolutionary growth for mankind.

It is here only, on this particular planet, that the segment which was teaching this area of experience became entrapped or enthralled with the actual sensation of the learning process; or the struggle, as it may be seen. As a matter of fact, there even are some who feel that by entering into, or lowering one's self down into, the element where the struggle or the conflict exists, he or she can better lift that area of consciousness or debate. But this never has proven to be truly effectual.

For there is a separation between the teacher and the student, there is a separation between the older and the younger consciousness or experience or experimentation. Therefore, entering down into it and experiencing it with the student does not allow for the teacher to have the mastery or control over and above it. In this had Jesus given you the admonition: be in the world, but not of the world. So, you can see that if you can be in the situation but not become part of it you have a much better control and do not lose your perspective.

THE FOUR ELEMENTS

All this is given that you may come to understand that the sensations of the physical embodiment and the life particular as it expresses on the physical are of one element of your experience, but not the total element of your experience. The way it can be controlled and evolved from out of its present miseries is by this aspect of looking frankly and clearly at the past and the errors of the past, but still rising above the past and the present into the eternalness of your spiritual Self, which is the God-given essence and substance from within; or your light consciousness, the fire of God.

Naturally, out of the water and earth elements you have helped solidify a consciousness or material substance that has become more apparent to your sensations than the more refined airlike level of your astral bodies, or the firelike substance which is the inner zeal and light of God from which you derive your energy; and all consciousness ensues from that.

So, it has been this devolution and this evolutionary process of the four elements which you manipulate and work with upon the Earth at this present time. The etheric substance, which is that which is beyond the fire element of your own individualized consciousness, is that which you are striving to reach at all times. This is the ethereal body or the electric body, or that which is the essence of God from within.

But the first step down from that element or area of consciousness is the fire of your own individualized light body or consciousness, which is the Christ Self of which we now speak and which you are striving to regain and to remain in. You have not lost the Christ body, or the light body. You just have lost your perspective and the total attention in it and with it.

EARTH-ASTRAL CYCLE

Once more man looks up from out of the mire of his earth and water existence, or material body, which is forever turning and changing and dissolving or re-forming from one substance to another or one form to another—dust unto dust, as in the truism of biblical scripture—and is returning again in a temporary period of time to that astral form which is the memory block or storehouse, and has within himself the inability to uninvolve himself from this constant wheel of remembrance, regret; reenjoyment, to a certain extent.

In the memory box or in the astral form of his being, when he does return to the astral planes around the Earth as he grows from one embodiment to another, he has the reliving experience. But he still has forgotten that he can recharge and change this air or level of memory into its pure substance and light, from which he has been derived, by attention to the light body or the consciousness of the God Self within.

He finds it even more difficult to step from astral consciousness into superconscious or light-body embodiment than he does from material matter or physical-life embodiment into the astral. So, for as many as twenty-six million years, as far as the Earth planet is concerned, those who became entrapped in this cycle of evolution have been moving from astral into Earth, and Earth into astral, evolvements and developments.

Many who are able to grasp the higher concepts have been allowed to go into other planets and planes and areas of development between incarnations. So, there is a great deal of recall now coming into prominence about the life on other spheres and in the astral areas connected with those various planes and planets in this solar system and, in a few cases, even beyond this solar system. Therefore, the souls that now are evolving through this Earth matter and this Earthly karma have become aware of some of the episodes that have taken them into other areas or schoolrooms or specialized lessons of technical substance.

ACCUMULATING KNOWLEDGE FOR MASTERY

You have many areas of learning and you have many technical jargons to comprehend. In the higher aspect of your soul mission you are required to have this consciousness and concept much within your grasp. To continue to desire this is a very good thing and is greatly encouraged by all of us who are working with the evolution of man throughout this entire solar system.

Do not take this as any kind of escapism on your part, for this could not be true. It is all part of the learning process and the mastery over substance or matter as it comes into its various forms and frequencies from out of the divine substance or essence which is God or Spirit Itself.

Never lose track of the major thought that is projected here. Man is to be master over all these things and man must have experience in all these things. For, as man is part of God, he is to experience as much of God's creation and expression as is possible throughout eternity.

Since he never can be God Himself and since he only can participate in the essence of God to the degree that he has consciousness of it and since God has as many means of expressing Himself and His variations as ever can be comprehended—and then, as soon as they are comprehended, a new comprehension comes into effect—man is obligated to, or is interested in, participating in these varied kingdoms and essences and creations. Let this be the primary motivating factor in all these lessons and truths that are being revealed throughout all time to you.

As man has gone from place to place he has accumulated all this knowledge. This knowledge not only is stored in the subconscious track for his own ongoing and ability to comprehend them, but the higher good and the true essence of righteousness and love are embedded in the superconscious or in the abilities of the presences of the light within to work with them, and to resolve them, throughout eternity.

UNIFICATION OF INDIVIDUALIZED SOULS

Here you have the individualization of each one who is a child of God; or rather, a specialization of his own identity as part of God. No two are alike. As even in the smallest elements of the material kingdom no two atoms or no two created forms ever appear to be exactly alike, how could you expect that the sons of God—who are so complex and have had so many individualized, personalized experiences—would evolve into the conscious state of being an exact replica of one another? This is totally false and foolish. We enjoy and benefit from the individualizations or the personality evolvements of each one, and each one supplements the other in these collective societies or governing rulerships or bodies of control.

So, you can understand for the moment that that which we call the society or governing command of a system is made up of diverse and many different types of spiritual sons or Christed ones who can benefit each other with their particular substances and their particular episodes of experience, then complement and supplement each one's overseership and rule of command so they can enter into one another or each other's dominion at will. For in the higher plane of existence, as already has been brought to your attention, from the light-body consciousness you can enter into anything else that exists, regardless of the level of its development, since all are parts of the substance and essence of God.

If a master or a light body of consciousness can enter into the smallest element on the lowest frequency development in order to enjoy or to experience it, then how much more so could those of light substance, energy and consciousness enter into each other's dominion and experience one another's vibrational frequency as a unification participation.

This is the essence of creation, this is the essence of reunification, this is the essence of brotherhood and part of the love development in the higher planes. You enter into or unify with that area or that individualization for the sake of supplementing and complementing him or her. But in the lower elements of contact, which you might call sexual relationships on the lowest plane of development here upon the Earth, there are motives and enjoyments which are far below this conscious area of supplementation and complementation; and are ones which do not subsidize, and aid in, the growth process.

But where you have a higher understanding you are reaching up toward this spiritual consciousness, and will do so more and more, so that the unification and the coming together of diverse or different souls will have a higher meaning and a higher evolutionary pattern than has been allowed for the last twenty-six million years upon the Earth planet. For the coming together of the various souls has been for the perpetration of the same error conditions and the same fears and negations that have existed ever since the entrapment of man into this dimension, and the fall of man as you now can relate to it in your own recorded history and progress.

COLLECTIVE OVERCOMING

But let us supersede all these things by dwelling upon the Christ or the I Am Self. For only in this area of consciousness are you going to be able to dissolve all the error mischief that has been created in the astral or the soul consciousness of the individual and of the race as a whole.

As all these things come to the surface, the only way you as individuals or as collective societies can grapple with them and handle them satisfactorily will be in individual prayer and meditation, and in group prayer and meditation, concerning these matters.

Then on the physical level you will have to have the strength and the courage to overcome this overpowering sensation of the past, and the memory and the error development, in order to reach the higher state of being which is your I Am consciousness or the light-body function.

Again the sons of God or the Elder race, as you once were and as you are capable of becoming again in the evolutionary process, will control the entire area that is known as Earth planet in its present relationship to this solar system with which we all are involved.

Now come the collective consciousnesses of those on Earth working toward this goal. We will see many happenings in this respect and many episodes of collective reasoning for the higher good of the whole. For the Earth is impressed now with these thoughts and with these patterns, not only by the fact that all are being channeled to this recording and this information but because there are many who have received it from within their own consciousness and have accepted it as truth and know it to be that which is of the higher essence and spiritual formula that now must come to pass upon the planet Earth.

You have been given as much as can be comprehended for the present. Therefore, I shall turn this channeling into a rebroadcast so the air or the soul essence of your memory track can carry it far and wide and can reach all others who are so embedded with it and so engrossed with the conflicts that exist within the soul level or the air-light of their being.

My love, my blessings and my truth I bequeath unto all. For as I make this manifest in and through this channelship, all shall be benefitted. Since this is the method by which all good can be overshadowing all that is of error, you likewise may proceed on this example and do it individually and collectively. I am Uriel. Love and blessings to all.

9. SOUL

SOUL AS RECORD KEEPER

Before expanding further on these matters, I, Lord Uriel, who returns unto this channelship during this discourse, will explain in further detail the formation, and the necessity of formation, of the astral body and its many functions in the service to the I Am consciousness or spiritual essence from within.

You are in accord with the record-keeping policy of all matters concerning history and evolvement of many subjects. It is from this essential policy or law divine that you retain this memory pattern and have been a race of souls who create records, bury them and burn them according to the policies and the prominence of that which takes place from step to step and development unto development. Therefore, in this divine creation or manifestation you have received one of the essential aspects of creating a responsible government for yourselves as individuals and for yourselves as a race conglomerate.

Let us assume that the individual was without any record-keeping device. He or she then would operate totally on impulse and intuition as of the moment, without any subconscious track to record, or to act as a brake upon, that present experience or episode. Although you are evolved to the sense of oneness and spiritual discipline from within, regarding the Godhead Creative Force or your instamatic creation from out of, or relationship to, the Godhead from which you originated, you still would not be able to conceive, or to recall, those matters which had gone on within others of your same ilk or yourself at varied times within the many episodes or experiences of your ongoing.

Therefore, regardless of how evolved you become and how many experiences you resolve and to what level you achieve the God consciousness in respect to your place and position in the kingdom of God, you still must maintain a record or a soul part. What is in that record and soul and how closely associated and aligned it is with the God essence or factor are entirely up to you.

SOUL COUNTERPART

You do not have to have a conscious part or principal in every single kingdom. But you do have to have a soul counterpart. In one aspect, it is the repeating of cycle, or imitation of divine principle in action. Since Godhead is a duality, a physical aspect of Father-Mother principle or positive-negative polarity in action, you, in the individualized Self state or I Am consciousness, substitute this polarity principle with a soul counterpart. Your own soul is what we are referring to here.

That soul counterpart is the subconscious side or self, the soul record or keeper of all which occurs in your ever going forward and spiritual development, and deeds which you perform in the name of the Father-Mother God Which is the Creator of all things and all kingdoms. So, where you are and where you go are in juxtaposition with a soul counterpart.

The soul-mate aspect often has been confused with this comprehension and this essential factor. We will not get into that subject at great length, but it is possible to give you some idea of how this error crept into the recordings of man and the race of man upon the planet, by saying that through this concept of having entirely and forever a counterpart to one's higher Self, a soul mate or a materialized form of soul as an adhesive part and an unbreakable impact within the concept and control of higher Self energy, it became a lesser or a more materialistic role to convey to the mind or to the personality on the physical level that a soul mate was the other half of your being, and therefore it was necessary to become solidified or bonded with that half that was missing from the beginning. But this is total error. This has no regard to what I am speaking of here as your soul mate.

The mating part is the marriage or unification of the I Am consciousness with the soul record or keeper from within your own subconscious tracks, in order that you become whole and carry on as a duality, a feminine-masculine polarity or a negative-positive polarity in relationship to the negative-positive polarity of the Mother-Father God Which is the Creator of all things. Therefore, you as the Son must be as an imitator of, or a minor symphony within, the major part or key to this entire universal structure.

RECORDS OF RELATIONSHIPS

In the soul record keeping are the aspects, the consciousnesses, the functions, the responsibilities and the learnings of all others with whom you are related in any form. You are related to some via the sex concept, or the fact that you utilized one sex or another in an embodiment and have used both in different embodiments. You are related to others in the sense that you participate in the society, nation, religion or different races or creeds that spring up in the ever-evolving participation of life upon a planet. You also are related to the sons of God via the higher manifestations or places of abode beyond this system of Earth, and the solar system of which Earth is only one single part.

Therefore, within the soul record there is the ability to call upon all these experiences or episodes. In other words, there is a library of information and deed by which you can recall or memorize some of the varied or related episodes in your own evolving grace and participation in growth as an individualized part of the whole.

Therefore, again like this great symphony of life, if you are but one instrument you may at times play at one theme or another, or you may echo a part of a theme, in order to play within the symphonic range of that which is being evolved or developed in the master music which is being screened and portrayed. It is ever thus through all evolution, and it is ever thus in other forms of life. It is this that must become a part of your understanding of, and relationship to, the varied forms and races in the solar system with which you will come in contact from this time forward.

DEVIC CONTACTS

In the case of the many kingdoms upon the Earth—animal, vegetable and mineral kingdoms—they likewise have their own counterparts or soul records. These are the auric manifestations or devic forces around which many who are psychic and sensitive can see beyond the physical form, or the natural phenomenon which is the form that is sensed with the perceptive abilities.

When an individual or a sensitive comes in contact with the deva of that particular form, be it a tree or an animal or a part of the mineral kingdoms of Earth, he is coming into contact with the entire racial structure or species structure of that particular individualization of itself. This is not to say there is only one deva for each species. But it is to say that within each species there is a grouping; and over that grouping of that species, which is much more complex than we can go into at this time, you come in contact with the one in charge, or the soul record to which that one or that individualization is connected. Therefore, you may read, sense or know many things concerning that entire aspect of that particular species.

It is through these contacts of the devic forces that you also grow and have a higher regard and relationship, and can have mastery over and governorship, toward those who are in those lower kingdoms. For, after all, we have stated and we reiterate that man is in charge of these lesser kingdoms in order for him to learn the control of his environment and all life form that is connected to that particular environment through which he is gaining higher understanding, higher mastership, and control of himself and that with which he is part. You have been given these keys and clues in so many different ways throughout time immemorial, and from all recorded history and from those who ever have been intuitive toward the higher aspects of life.

So, within these different species and their counterparts or devic controls you have the record of the entire race of man upon the planet. For the devas are in conscious awareness of, and are intuned with, all the subconscious activities of the race of man, particularly that particular segment of man which is in direct contact with that species at that particular time.

ASTRAL INFLUENCES

To be specific, a certain race or society of man in a certain geographic area would unconsciously be controlling the geographic substance, through the devic contact in the place where he resides. You have been given information, for example, that your American Indians had a great deal of contact with the very environment —their land, the weather, the animals and the mineral kingdoms— around which or through which they had their growth process and their dominion. Always, in the higher aspect of this type of primitive society, you had those who were in conscious contact, through the subconscious mind, with those forces and who worked in conjunction with those forces for proper environment for the people they served at that time.

This would continue, naturally, when those who had great power and understanding transferred from the soul evolution into an astral body and no longer were equipped with a physical vehicle through which they served their particular race or society. You have been told, and many who are psychic sensitives or prophets of the present day have seen and have known, of those in the astral kingdoms still in command over the vegetation, the animal life and the mineral kingdoms of the planet in their astral consciousnesses and in their astral forms.

You have too, within the societies of some primitive religions—

such as the American Indians, and the African primitives as well, and those in the Polynesian and higher guarantees of Eastern lore —that their contact is with the astral entities over their present embodiments, over their present environment and over their particular growth patterns. They still are relating to their astral ancestors for control and for contact and for assistance.

This is because they have not been taught the higher aspect of the I Am consciousness, and the superconscious role within the race development. Because of this we have had teachings which have been brought to all these places and regions in order to break this contact between physical and astral associations and interplane functioning.

IMPORTANCE OF SOUL RECORD

But it still is not sufficient if, in the I Am consciousness and control, you completely wipe out or nullify the effect of astral influences and astral entities who can and who do work with, and are connected to, the entire soul evolution of mankind upon the Earth and in every other plane of existence throughout this solar system and beyond.

You must always think in terms of this subconscious recording. You must always think in terms of this soul-mating within yourself. You must never lose sight of all these episodes, learnings and record keepings, for your own information and growth and evolvement. So, although you take the most highly developed spirit in your evolutionary conveyance, such as mastery and Christed beingness, you still must have within your concept the ability to utilize, and to be wedded to, that which is the best within the soul or subconscious area of growth.

Those who are in total mastery of the I Am Self or light-body functioning upon any plane, planet or dimension are those who have cleansed sufficiently the soul recordings of themselves, and have cleansed within themselves the concepts which the race has portrayed or has placed within that soul record keeping. In other words, or very simply put, these masters are ones who have cleaned up the library records of their own soul messages. For within the soul aspect or subconscious record you are receiving impulses or ideas or thought patterns, at all times, coming from those who are in the like position or are in similar juxtaposition with you in soul or spiritual evolution, wherever you are.

You never are beyond this state of limiting yourself, never are isolated from all others who are in the same area of growth and expression. You are forever in relationship to all others of creation within that pattern or growth or development. You never lose your Self, your higher Self, to the expanse of God creativity and become so absorbed in that God Creative Principle that you lose touch with those who are in the growth pattern or evolutionary pattern with you.

So, it is important that the soul pattern or record-keeping function is ever intact and in communication with you as a spiritual being evolving and growing in further experience, exploration and in divine grace and function. Therefore, you must be grateful unto this soul or negative polarity position, and respect its function and its part in the playground of experience, and give it its proper due, and become one with it but in control over it.

This also has been spoken of as the husband and wife relationship, in the scriptures, wherein one has said the husband or the I Am consciousness must have control over the wife, and the wife may not have control within the house; the house being your own vehicle of development, your own vehicle of expression. The vehicle of which we are speaking here would be the I Am Self, or the light-body expression.

But, at the same time, the soul must be cleansed sufficiently so that it is in complete rapport with, and in complete agreement with, all things which would be of the divine; and would not permit any of the lesser forces or any of the lesser ideas to be projected in from lower kingdoms, lower desires with which it is attached in the race consciousness and which therefore can interfere with the decisions of, and the dedications of, the I Am Self consciousness to perform its duties as an individualization and identification of the God Self; which is the highest form of being and impersonalness, but yet personal.

RESPONSIBILITY IN FREEWILL CHOICES

You have been afforded a great deal of time and energy along these paths in the eons of growth and understanding and unfoldment of information. You likewise have fallen into the traps of many misconceptions and of many misinterpretations. However, if you as individuals with free will, and in growing toward this absolute concept, would not have the opportunity to explore them and to utilize them and to find them in error, you never would have the control or the conscious application which is yours to perform for the higher Self and the light function of God-being.

Many do not understand this, and have questioned it and are ever berating the system that God has put into motion. It is all good and well to berate the system and to fight the system; but, since the system is put into motion by the God Force Which is, and is the eternal substance and essence of all things that are created in order and in harmony, there is just no sense to fighting that system.

For that system will outlive all things that argue against it and fight in the attempt to change it. It never can be changed; it is. Therefore, you will do a great deal more for yourself by cooperating with that which is than in fighting it and trying to change it. For although you succeed in a temporary measure in your own minds, this is not relevant, nor is it practical. In the end you have to move away from that wrong concept and to resolve yourself to devolving from it; in other words, falling back through the trap into which you have fallen, and rising up through the very doorways or pathways you yourself have created in this mistaken identity or concept.

This is another major point that has to be made in this soul aspect, and the relationship of the soul to the spiritual I Am consciousness. Whenever there has been a wrong road created by man—or any other creature or element that can force a pathway into divine energy, matter and form—you must travel that pathway until you see the error of its end results. When you come into conscious understanding of the error which has been created, you must backtrack along the exact same pathway in order to arrive at the crossroad where you had been given the choice. This also is part of the freewill aspect.

In other words, if you are at a point in consciousness and development where you can take one path or you see another that may be more promising, though it is not, you are forced to retravel or to reestablish your steps along that pathway, undoing as you go each step of the way that which you have created in the process of developing that path. When you reach the crossroads again, the revelation comes to you as though for the first time.

This is the purpose for that. Your subconscious mind must have the unwinding of the threads or patterns and be reestablished according to proper and higher concepts. The subconscious mind, which is the record keeper, at all times must have the total record in a very definite and simplified method. You cannot jump from one area into another without going through the varied steps you yourself had created in order to come to a certain area of concept or conclusion.

It is a very simple matter; and must be taken into deep consideration for this purpose, also. Unless you are willing to take the responsibility for what you create, you may not be given the higher functions and responsibilities as a child of God or a co-creator with divine source, energy and supply. Therefore, by learning this process through the subconscious record, all is impressed, and deeply implanted, in the record-keeping device which we call soul.

SOUL SUBSTANCE

The soul aspect of even the highest forms of creation is ever intact; it is never loosened. But you take upon yourself the soul substance from that area, and the elements within that area, of the place you have chosen to have your deepest and most rewarding experiences. Therefore, the soul substance of one type of creation is not always exactly the same as the soul substance of another.

As stated before, the soul substance of a rock is not the same as the soul substance and record-keeping device of those who are the sons of God. However, within the race element of God's sons or co-creators He has inbuilt in them the similar substances or elements which are complementary and coincide with that in any place, era or time that such have had experience and exploration.

Therefore, you can relate to all who are in the Sonship relationship with God anywhere, any time, anyplace throughout eternity. It is so created so that all may become one within the conscious application of that oneness; in other words, that you may know your oneness and supply your speciality as a service to your fellowmen, and know the brotherhood of the race and the Sonship relationship rather than finding yourself unique and separated from that relationship or that Creative Force Which is God Self within.

SPIRITUAL ESSENCE CONTACTS

You are ever in contact, through the spiritual essence or the I Am consciousness, with every other form of creation; but not necessarily on the soul level. Therefore, there are many in the soul level or subconscious relationship who are upon the Earth at this present time and can see absolutely no relationship, and find no rapport, with the elements or the lower kingdoms.

This is because they have not raised themselves, or have not eliminated from themselves the blindness of their soul or subconscious reportings, and are trying to relate to these lesser kingdoms or individualizations via that aspect instead of through the personalized I Am Self, the substance within, the essence of God or Spirit Which is in all things and Which has created all things and never can be eliminated or wiped out of any form or force of creation.

Again you all are reminded that in coming to contact and recording with those who are of other elements, those who are of other kingdoms, you may do so through the I Am consciousness or the spiritual aspect of yourself rather than through the soul substance, which may have within it certain blockages, memories and patterns that are not suitable to that kind of relationship. But as you do this you create within the soul substance somewhat of a new pattern and a new guideline by which you can have these relationships and enjoy the rapport and exchange of energy with everything that ever is created.

For this is also the goal of mankind: to have within his experience an interrelationship, a higher form of exchange, with everything that exists; so that everything serves the One, and the One Which is in all things is being served and sustained and rewarded. You thus are fulfilling that which you have come to do, which is to be as a child of God. Since God or Spirit exchanges within Itself all Its forms and all the lesser kingdoms without prejudice, then does it not behoove the sons of God, which are the imitators or co-creators with God in His aspect of trinity, to have a similar type of relationship with all form, with all matter, with all Creative Principle?

CLEANSING SOUL RECORDS

In this I have given as much of the soul participation as is necessary for a preliminary discussion along these lines. In the evolving process of soul betterment we must think always in terms of the cleansing process and of the reestablishment of a higher functioning for the soul part of man. Because we are evolving up out of the matter or personality self and are coming into the spiritual aspect or fourth dimensional frequency light-body manifestation, you can cleanse and release all these soul impacts via a new and higher routine.

But let us not dismiss the lesser methods until the higher ones are totally understood and can be properly utilized. In the release of soul impacts, see always that the light of the higher Self, the Divine Energy Which is ever perfect, can replace every imperfect thought-pattern and impulse that have been planted there; and can be restated in the intention for God Self in action, rather than man in action.

In this aspect, Jesus of Nazareth, who is the master of this procedure and process upon the Earth, said: I am the son of man, but I become the son of God. For in his Sonship and in his Christed realm of understanding he realized that the soul process of cleansing from a man-made level, and his relationship with his fellowmen in that area of growth, could and must be replaced by a higher Self divine guidance or Sonship, and that it was not a transfer or a superstructure replacement but a gradual understanding of one level replacing a higher level.

He was not trying to eliminate his soul record, nor his soul memory nor his soul pattern, which still exist and can be utilized. We will try to explain how these various soul patterns or soul memories can be extended out for a sole purpose of bringing about a lesson and individualized instruction, rather than going through the many more-devious means of expressing and teaching the same principle at different times.

PROJECTING PRIOR INCARNATIONS

Here is the routine in this respect. Within the soul impact or impression are many experiences or embodiments throughout eternity. When one has reached spiritual I Am consciousness or Christedness he can call upon any one of these soul impacts or soul thought-forms and can project it out for an immediate lesson or an individualized thought gradation. It serves a purpose for the time and it serves an impact on those who would receive it.

In the transfiguration on the mount when Jesus projected two incarnations of himself, Moses and Elias [Elijah], as well as his higher form as Jesus of Nazareth, the disciples witnessing this were able to see three projections of the same spirit on three separate occasions of soul evolvement. They were able to comprehend that the three were as one, but the three served different purposes at three distinct eras and evolutionary periods within the growth pattern of that particular spirit or I Am Self consciousness.

It has become necessary in these end days of time for many to experience the various extensions or soul patterns of themselves at different parts of their unfolding. In order to do this they must come to the comprehension that all is under the control of the spirit or the I Am Self incarnate. In this period of comprehension one can see or receive simultaneously several soul incarnations or patterns at a split second and can conceive of their purpose, of their interrelationship and of their various aspects.

Again you can liken it unto a symphony orchestra, whereby each instrument may be playing its own particular theme for the moment, and all together make a symphonic whole or total structure; not in opposition to that but in harmony with one another at that time.

This is absolutely necessary to the Christed Self construction. For in the I Am state, or the light-body state, of evolution man will regain his knowledge of his many places of residence and of his many aspects of learning within his soul recognition, and thereby will be able to function in them simultaneously. This shall become very, very important. But those who are not in rapport with, or brought into alignment with, their soul part will have tremendous confusions and conflicts from within.

UNION WITH SOUL

Let us discuss the race structure and see within it that it is necessary that the race at all times be in awareness of this higher understanding and this higher teaching, so that as the race evolves from third into fourth dimensional frequency the race once more can appreciate and can utilize that halfway step which is the soul embodiment or the soul frequency or the subconscious memory pattern that is the server for both the spiritual I Am consciousness or the divinity within and the personality or mortal aspect of the physical incarnation or body that is being projected for the temporary period of time which we call life expression.

You have been given this in order that you begin to have union with that soul aspect within yourself, within the race of which you are a part, and within all other kingdoms that are upon the Earth; which shall reach you from the outer reaches of the universe in order to help you lift, grow and expand beyond your present level of development.

For of utmost concern to all of us in the hierarchal structure of development and control and governorship is the fact that man is evolving as an entire race, with the planet, from a third frequency form into a fourth frequency form or fourth dimensional state of consciousness. He again will relearn and will reestablish himself in the light-body function, as was the original intent and the original goal for his expressing upon the Earth, through the Earth, and forming the Earth as a proper school and residence for his development and for his greater control within the kingdoms and the many mansions available to the sons of God, who are the children of light and the enlightened ones of wonder.

Let it be so in you. Let it be formulated and frankly impressed upon your conscious mind so the cooperation can begin from within the I Am Self to the outer consciousness; and return from the outer consciousness, where you are now in the personality self, back through the soul and up into the I Am state and grace of life. So be it in truth. This is all for the present. I am Uriel.

10. GOOD AND EVIL

MAN CREATED EVIL

It is on this basic question of good and evil that all progress and regression have been fought, have been struggled with and have been won. In order to resolve this basic and different approach to that which has existed in man from the beginning of his search for freedom, and at the same time for devotion to his God Father Creator, I address my people and children of Earth. This is Sananda, Christ Jesus of Nazareth in last embodiment on the Earth.

You have been bothered with this fundamental problem and have fought many good rounds in the quest for truth concerning it. There lives not a single soul who has not quested the rightful answer to this in his soul evolution and in the battleground of life eternal. So, let us now grasp a few concepts and make them steadfast in our eternal determination to give righteous and loving proof of the fundamental laws within this universe from which we are created and out of which we have our sustenance and existence.

Let us begin first with the scriptural reference of the creation of all that may exist. In the beginning was the Divine Creator. Before this there was always the Divine Creator, or Principle Itself. So there never was a time, never can be a time, when this is not the fundamental reasoning and thought power.

Out of this thought power and mind which is God, all has been created and has been sustained and has been put into motion, for good and evil. But all things were created before the evil even was mentioned. That is the sum and total of the entire question. It has not been the God Principle or life substance which created evil; but man, who contemplated the creation, who thought in terms of error or evil.

From out of these thought compounds he made himself a master unto them and he became subjected to that master which he himself created. So it is god-man who created the force of evil in his thinking, and not God the Father-Mother Creator, who possibly could exist in the same world or in the same consciousness as such a concept.

MAN AGAINST GOD'S PURPOSES

Let us start from this premise and see how it resolves itself. Out of all the good that God created, He put man in charge of these kingdoms; and He guarded them well with other creations which He made, from which god-man could have his relationships, learnings and protections. These are spoken of in the Adamic race upon this planet and in this solar system, through the parables given in the book of Genesis. All these things are related to other such scriptures in other such developments of the race of man in his conscious memory of the beginning of time and his place in that scheme of creation within the scope of time and space.

So, there never was a time and there never was a place wherein God did not exist and did not have a reason for His existence and a purpose behind all this. But in man's search to go against the principles or to interpret these principles or to taste of certain ones of these principles and life forces that were not for his readiness at a specific point in growth or in evolution, he became a deceiver and he became a negligent keeper of the flame of life and the truth and the purpose of God's force in all creation.

This was the beginning and the downfall of man; not any one creature, not any one series of events, but the entire means of doing that which is not for his highest advancement and good, according to man's idea of how the progress should be made and wherein he might dwell and take up certain roles and experiences. This is all outlined in a brief and succinct manner via scriptural references, parables and tales that come out of the consciousness of man in the memory pattern or soul relationship of his subconscious record-keeping machinery.

CREATING AND EVOLVING OF ERROR

Let us not deceive ourselves that such conditions are only a matter of mental projections. For man is the Son of God. Under this regulation his mind power is very capable of creating a substance and matter out of his mind powers. So, by creating ideas and having thoughts independent of the role, the mission, the plan or the scheme of proper evolutionary progress and growth, he has created certain levels of being, certain creatures of being and certain deeds of being that are considered evil or erroneous to his proper growth and progress.

All this has been termed satanic or of Luciferian force. But it is not existent, in the sense of God being in existence throughout eternity and without fault. It does exist in the sense of man's creating or helping to create a force or energy projection that, in the scheme of divine law, must have its extent or playout in the drama of cause and effect, or karma.

As has been stated before, nothing can exist out of divine law, order and harmony. Therefore, when such force is created, it begins to evolve of itself and is able to create out of that force of energy and to become masterful and predominant, if one allows himself to dwell upon this condition or to see or to give credence to it. Therefore, the error or evil that men do creates a further situation which is compounded by that and enlarges upon it and begets of itself further deeds, conscious contemplations and projections which then become mightier than the original concept or thought or projection.

In all these creative facilities, man has impressed upon himself and his environment these evils or errors. In many other creatures around him he has implanted such error thoughts and conditions, as well. For he has lived with the beast in the field and amongst the rocks and trees of the field and has seen of them certain conditions which now have become also error or evil, in his way of thinking.

All that is right and true and progressive and developing is of God. In God's laws and methods there are life force, energy, growth; and a period of deterioration and dissolution when it becomes unto itself nothing but substance matter; and then re-forms and reenergizes into another or higher aspect of itself. For energy can never be lost; it is ever continuing. This is the rightful progress.

Out of this, error conditions also can be made to have a consciousness or a process of development, in this way: when they can destroy, hurt or harm and have a power over the ones who see it as a power in itself, they take on themselves a certain self-propelling and self-energizing power. This has been the ability of man, in his thought projections, to give credence or power unto that which is of a destructive nature.

DESTRUCTION AND EVOLUTION

But destruction, of itself, is not evil. For, as I have attempted to point out here, the process of growth and evolution is a period of dissolution or consumption and re-formation. Therefore, the development stage of destruction, in a constructive process of growth, is not in itself evil or error.

But man, seeing this process, partly understanding this process, has leaned toward this process and has named it evil. When he has done this in any state of consciousness, he gives it the power only of its destruction and sees not its constructive purposes. Out of this he has created the evil or the error; or the devil, that he has called it.

Because of this, certain forms of life have not been able to progress according to their highest instruction, mission or plan. There is a pattern for every form of life, which is to give of itself for a higher form of life or to participate in the exchange of energy or frequency with another of a similar type of energy and frequency structure. But where this process is only destructive and not constructive it has become as a terror upon the Earth and in the minds of many. This has become the devolution or the destruction of matter as it exists.

However, in the final analysis and in the return unto energy force, it has to have its own personal reevaluation. In other words, when it has been destroyed totally as a species of destruction, then it must reevaluate in its own soul structure or astral form that which has preceded, even if it has taken eons of time.

RE-FORMATION OF ERROR FORCE

Let us examine a type of species upon the Earth which is called parasitical. If it has a period of time in which it has not added to, or aided in, the construction of life form in that area of development, eventually it is totally eliminated or consumed and brought back unto its original state of power, which is the essence or spiritual flow of life. In the soul or akashic record of it, it becomes reevaluated, re-formed and restituted in another place at another time for another go-around in the cycle of evolvement and interrelationships.

Only because man in his fall has accepted such has it been able to exist in his environment. If man, in his state of grace and consciousness of aiding all such instruments of Spirit, would give them all the conscious appreciation and assistance to become co-inhabitors of a state of grace or a state of residence wherein they all are arising in evolution, he would give that individual species or that form the opportunity to grow as he is growing; even though sometimes he appears to be going backward or falling into more disgrace.

From these comments it is not to be understood or taken, in any manner whatsoever, that error or evil cannot exist, even if it is seeming to be harmful. For there are areas of consciousness within mankind himself which refuse to accept the God essence or spiritual inflow of eventual good and eventual restitution as a son of God. Therefore, that power and energy force have to be broken down totally within the concept or the sociological structure of the race in order to be restituted or re-formed out of mind substance and matter.

This is done by the saviors, so-called, or by the assistants to those sons of God who are the protectors of the light body or the spiritual essence of mankind known as the angelic force and kingdoms. These usually come under the archangels, who are in charge of all things that come out of the race of man in relationship to his God Self and his eventual status as a Son of God and a governor or ruler over all lesser kingdoms; as also spoken of and indicated in scriptural references, particularly in Genesis.

THE GODLIKE

This brings about another point of discussion here in referring to the gods and goddesses of legend and myth, and the various mystical stories that have arisen out of men's minds in order to explain the progress and evolutionary grace in this dominion called Earth and this solar system. It has to be understood from the beginning that the term is made only in reference to those who have a Godlike relationship or a Godlike reference in their beingness and in their own status, not to ones who are a hierarchy of gods and goddesses. For there are no such things as gods and goddesses in the sense that God shares His kingdom or His areas of control.

God is law. God is all the attributes and divine principles that exist throughout all creations. God is not an individualized soul or personal aspect of being, but that One Which is all that can be, all that ever has been and all that ever shall be, and exists in all forms of creation.

Those who come to this consciousness, those who have this concept, those who can enact these principles and teach them and demonstrate them are as gods, or Godlike. This is the frame of reference in all such literature to which we refer.

This is not the same as the angelic kingdom, which is a species unto itself and which comes under the protection of the Godlike law or the God-made laws of the universe in order to enhance and to protect Itself from Its own creations; namely that of mankind, which is the Son of God or the manifested form through which all the principles and laws of God are being enacted and are acting themselves out and are being unfolded for further and ever-progressive developments.

GOD CREATES ONLY GOOD

In the terms of the goodness of God, herein is the statement or the definition that indicates that all which God does is from goodness and can never be from evil or error; because all that God ever is, has done and will do is for progress and unfoldment and further development and experience. This in itself, regardless of suffering and defeating experiences, is the proof or the reason for constant evolutionary matters to be undertaken and to be experienced properly.

Man must come to realize that even his worst experiences, his most defeating blows are those which eventually can be for his highest good and his greatest understanding and for his best demonstrations eventually in his growth upward toward his proper relationship with God.

Since originally and in the beginning man is this part or is created out of this substance and energy, and fell into this evil or error way of concept, he himself brought about his own destruction, and not God. For God did not create two energies or two concepts or two plans. God created only one plan, which is the good plan of evolution, of experience, of further unfoldment toward Himself and toward all of His creation and toward all of His myriad ways of expressing Himself. Through himself man then can experience all these varied aspects of God.

EVIL IN PURSUIT OF MAN

But whenever man has decided there is a different plan or a more self-satisfying plan or a more controlled plan that does not take God's plan into consideration and the other creations into consideration, and wishes to be self-sufficient or isolated from this condition, he has produced a second-rate energy or a second-rate series of events. He has put into effect a cause different than the original or primeval cause, Which is God.

This is the evil that men do which lives after them. This is the evil which follows man from generation to generation. It is of this evil that man has memory recall in his subconscious mind. It is this evil or devil that pursues him and has control over him. For it is man's creation, and man must undo his own creation step by step.

This is exactly what is now unfolding upon the Earth planet in the time of *now*. Every error condition that man has perpetrated in his development as a Son of God is being unleashed upon him in the memory recall. In the memory recall or soul consciousness it has had so much power from generation to generation and from mind to mind that actually it has taken on substance or matter.

But matter, in this sense, can be dissolved and returned to its original premise or to its original essence, which is spirit and pure mind energy. From that original state it can be re-created into its proper channel of good and right and truth and beauty. This is the obligation of man now upon the planet.

Is there any man upon the planet or in all creation who of himself is so evil and so filled with error guilts and conceptions that he cannot be restituted or saved? No, this is an impossibility. For that would deny the fact that his original concept and creation are that of essence or primary energy, which is God or the God Self within.

CAUTION IN CLEANSING ERROR

But scraping away all the effects from the original cause of his first mistake or error condition is a very detailed process, and sometimes a procedure so ugly and so painful that many cannot bear the brunt of it or face up to the consequences which they have unleashed. So, you must be very careful in the processing of this evil and error condition, not only in individuals but in whole segments of society.

In addition to that, when you get into the higher control over matter and the principles of life you will come into contact with the evils or errors of certain energy forces in the natural kingdoms as well. For there are some, in the conditions of man's life upon the planet, which can bring about very dire, destructive levels of generating force which can overwhelm and sweep away all the good that can be apparent to the immediate attention and mind for it in the present cycle of time.

There will be some of this cleansing taking place in the following century, when more light workers are embodied on the Earth and the control of these forces can be made in unison or collectively. It is not up to individuals, at this stage of their recollection, to unleash all of these atoms of hate and violence at the present time. So, in one sense you must be zealous and in another sense you must be cautious of that. You must understand the delicacy of your power in the first stages of realization, and not undertake too much cleansing of evil at one time.

DEMONSTRATE ONLY FOR HIGHEST GOOD

It is similar to the case or symbolism given in the story of the temptation in the wilderness. When each temptation was presented to me in conscious mind during the lifetime as Jesus of Nazareth, I dealt with it in its own level or as much as I had power and conscious control of. But the final and last temptation was to throw myself over the cliff of, or the narrow bridge of, conscious awareness; because, with the concepts and the powers with which I had become aware, I could not of myself be destroyed. This is absolutely true, for no son of God can ever be destroyed. But the body can be destroyed. It was there that the wisdom of the spiritual Self within said: I shall not tempt the Lord my God at that point.

You in your wisdom shall echo those thoughts and sentiments when you are confronted with a demonstration that is beyond your own ability to perform, or your own experience within the light form, at the time where you are recalling it or utilizing it for the highest good of all.

This must be the key and this must be the code by which you operate in the light body and in the Son consciousness: that which is for the highest good of all is that which you can demonstrate and will be proficient in. It is that which will back up your own attempts to be of the light and in the light and to show the light to others who are of a lesser understanding and enlightenment.

Where it is to demonstrate only the power for the sake of the power and the demonstration, you will not have a backup, in the spiritual sense. In other words, your own spiritual consciousness or your own I Am level of speed in light will not give you that energy to protect you and to enforce that which you set out to do. But where you are expected to demonstrate for the highest good of as many as can be aided in that demonstration, you will be given the proper energy and control over that force.

Let this always be in your minds: those of you who are coming into the light and shall have the power of the light or the body of the Christ within, showing on the outer, can demonstrate it only as far as it will do good for others and be for the highest evolutionary grace and graduation of your fellowmen who are entrapped in this kingdom of Earth and mortal consciousness.

Is this not our only goal? Has this not been the purpose for the Elder race and the brothers of the light to accept the descent into matter and to reeducate those who have become so entrapped, that they may rise up and out of the entrapment of one level of experience, where they find themselves in the third dimensional frequency form? It must be yours, in this dedication and desire.

For all else is of evil, or all else is of error. Out of that error condition you will continue to perpetrate certain error concepts and certain evil doings that will be reaped in another time and in another place. Error of itself is not eliminated just by the speaking of it. It must be gone through and reestablished and broken down into its original primal source, which is good or God substance. From there it can be rebuilt into the primary purpose for which it is unique and special.

EVIL CONTRARY TO DIVINE PLAN

As you have been told and as we repeat, all source is of One. Each combination of the Source that is One creates a form or an energy release that is a unique species, purpose, plan and project. If it goes against that particular purpose, plan or project it becomes evil or error. This can exist in thought, it can exist in mind, it can exist in form and it can exist for eons of time, until those of the light are protected enough and energized enough to break it down into its original substance or light-righteous matter and mind and make it restate its intention, form, plan and project.

Within the seed is the final fruit. So it is within each structure or essence or combination of life, form and matter. Within that are the plan and the program by which it can proceed upward and outward and become whole and resigned to its proper place in the scheme of evolution and development.

But should any other force, or should the force within it, deem it wise or expedient to re-create or to recharge that plan or program, it becomes of itself an error or mistake in matter. This is what we are overcoming and this is what we are erasing and this is what we are embattling at the present time upon the Earth planet.

You have been good because you are of God. You are light because from the essence of light and love you were created. But the error or evil that has come from this is of a temporary nature, regardless of its length of stay and the power it has emulated from out of this very goodness which has been bequeathed unto you.

LESSON FROM EVIL

No other form or level of creation has as much power to do evil or to make its own free way as has mankind. For no other form has been given the understanding of freedom and will and self-determination. But out of this self-determination and out of this will against the proper course of events comes a great lesson that shall be learned and shall be seared forever within the pact of man's own understanding and growth as a race, especially as it concerns this planet. And whatever concerns this planet will affect this entire solar system.

There will be many, many lessons we can perpetrate and can teach for others who are going in a similar direction or who may have the erroneous concept that there are paths other than the ones that were created for them in the exploration of their Sonship with God. So be it.

Let it begin in your own consciousness first. For that is where the battle will take place. Then let it extend out from that consciousness, in concerted action with those who are of like mind and concept and who can grasp this situation well in hand and do with it those things that shall speak well and good and right of the sons of God evolving from out of the form or life which they themselves created for themselves in error or evil.

I am Jesus of Nazareth, the Christ who came and who will return unto this planet in the days of good and will bring about all restitution of matter which has fallen into disrepute and disarray and disruption. Amen. So be it. It is truth.

11. BRAIN

CONTROL BY SPIRITUAL SELF

Nada, expressing in and through her vehicle as Yolanda of the Sun, a channel and disciple of the light. You have been given a number of very explicit terms, as concerns the evolution of man in and through his own vehicle physically. But you still do not terminate your prior conceptions, because that is much too difficult in the course of a few definitive remarks. So, we attempt to give evidence, and particularly some specifics, that can help in gradually evolving your conscious understanding. May it bring about a deeper reverence for that which is the spirit and the consciousness which goes with the spirit in conceiving of and performing its duties through a material substance.

Your brain is the most important part of this connection. It is in and through the brain circuits that man eventually can come to discern the incoming and outgoing spiritual flow and light-body manifestation. It is because of this that no man ever has seen the light body, in a physical part of the vessel. You never shall see it. For you may not probe those sections of the brain where the spirit controls the entire mechanism. It is in the interior sections of the brain, the first and essential cord that creates and connects matter with Spirit.

IMPROPER MANIPULATIONS

Your soul record also is related to the brain mechanism. By manipulating or extending certain functions of the brain faculties you will unleash many mysterious, marvelous methods for soul growth and recordings. But let it here be understood that none of the mechanical ways or none of the devious and self-perpetuating ways are considered proper by the spiritual Self. For all must be under the command of the spiritual Self. This is why no higher school of learning and spiritual teaching can approve of those methods such as hypnosis, drugs, mechanical manipulations via surgery, and other such techniques.

Yet has man experimented with all these things throughout the eons of time upon the planet. We know and understand their purposes. For in many places such have been used to perform a necessary deceptive requirement, that being the recording of these matters upon the Earth to prove their inability to be all-inclusive. There are some who would be accorded these advantages and still not develop according to the prescribed outline for them. This is because the higher Self of that individual did not wish, and would not permit, the evolvement or the revelation to take place.

For a short example, may I point out that hypnotism, as practiced in the Earth as mind controlling the varied subconscious part of the record mechanism, cannot work upon all, for there are those who are so evolved that they will not submit, and cannot release, the record unto any other but their own spiritual evaluation from within.

This is the type of example that forecloses on absolute statements and recommendations via scientific communities who wish to make flat and all-inclusive statements about everything concerning all conditions. But where spirit and souls are involved you cannot conclude scientifically any formula, for that is not in the province of science but under the dominion of the God Self within.

Since each individual is a unique pattern and has a unique pattern to perform and to fulfill, therefore a so-called scientific formula or set of circumstances never can, and never shall be able to, be effected upon the spiritual accesses. This too explains, in the minds of many, why certain spiritual functions such as telepathy, mind power over certain objects, prophecies and many such transmuting effects of spiritual force never can be controlled via limitations or comparable objectives in this way.

SCIENTIFIC STUDY

You have been given the understanding that Spirit connects all things through a magnetic force field. That is also true of the brain faculties or brain-wave matriculations. When you wish to experience a spiritual phenomenon through individuals so raised to that consciousness and cooperation and understanding, you will do it in your scientific laboratories under controlled conditions via magnetic wavelengths.

But many who are experimenting along these lines have yet to find the recording equipment sufficiently sensitive to pick up, or to substitute, their findings in the proper relationship with this field of endeavor. So, do not expect an immediate response to this prediction or factor at this time. But before this century is completed you will have scientific laboratories able to demonstrate what is mentioned here. Until that time you will work with electromagnetic force fields. This relates more to the soul or memory level, whereby dreams and thoughts and projections of thought forms can be recorded, to a certain degree.

But again, under this premise you must remember that only via the soul and spiritual consent of the parties involved can the demonstration be successful. This precludes any who attempt to put it on a strictly material or scientific basis. Regarding the word *scientific*, we mean it here to refer to those who specifically are seeking physical scientific evidence or laws that relate to the soul and the spirit. Never could we refer to those who can comprehend and relate the divine laws and spiritual sciences to this matter and subject.

But before we continue, you must regard all such subjects and experimentations as merely a part in an overall plan and program that shall enhance the speeding up of spiritual progress upon the planet. Until all who participate in these types of experimental projects are cooperating on the same wavelength or purpose, you cannot have the progress which you will desire and which can be relayed to you via mind matter and spiritual projections.

In this aspect we make most-specific promises, in this way: we could and we would help those scientists who wish to speed up the spiritual evolutionary process via the means outlined here, if those individuals truly were concerned with the overall spiritual upliftment of all men equally upon the planet and could comprehend and could appreciate the entire spiritual program as it relates to the Earth and this solar system wherein the race of man returns to his full potential as a Son of God within a certain frequency vibration.

BODIES OF MAN

Under these circumstances we will concern ourselves with some of the scientific facts involved. One, all matter operates at a specific vibratory rate. No two objects are exactly alike, even though they remain within the same orbit or body form. In this way you have different cellular structures even within the same body form.

Within the same body or entity you still have grades or levels of vibratory frequency, concerning the several areas of evolvement and development. First, you have a physical body, then you have an astral or soul body, then you have an etheric body. All these three levels are interrelated, interpenetrating and concerned with one another. One cannot evolve or move too far away from the proper, harmonious relationship with the other.

Naturally, the spiritual or etheric body is above and beyond any concern with the two lower bodies, meaning the physical and the astral relationships. But still that etheric body cannot manipulate fully, cannot sustain life or cannot give energy to the two lesser bodies unless there is total harmony with the etheric body coming from the physical and the astral bodies.

In other words, if the soul becomes so disharmonious that it diseases the physical body by its projections of unhappiness, fear or other conditions, then the spiritual body is obliged to withdraw its essence and life sustenance from that form which is the physical body.

Therefore, the physical never can control, or dictate to, the spiritual or etheric body. The soul body is the relay station between the two, so that it reflects both at the same time and equates the proper condition for that present evolvement and experience.

MIND CAPABILITIES

All these things are recorded and manipulated through the brain mechanism. Although much of what the brain is capable of doing is known and recognized by some to the present date, none have understood fully or have coordinated every facet of it. For there are areas of the brain that seem mysterious or unrewarding or unrequiting to the man-made laws and knowledges that have been set down for the present time.

Although sciences in the past have related to these things, they have not been passed down in a sufficiently acceptable form as far as present-day man is concerned. Although some civilizations in the distant past had a greater knowledge in some respects, they had a greater disregard for what was being accomplished than was suitable. So, all this knowledge was withdrawn from the race memory or conscious mind. But nothing is ever erased from the soul memory. That is the reason many of these things are coming into prominent discussion and discovery at the present time. It will be unfolded and it will be utilized properly.

By the brain wave or thought projections out of the material part of the brain, or the material-controlling aspects of the brain, we can see even today on certain equipment what the mind is capable of doing. But we have not explored sufficiently what the soul can do when awakened to itself, nor have we even begun to think in terms of what the spiritual Self—which also is connected, through the brain, the central cord and the deepest recess area in the brain, with all the other parts of the soul and body functions—can express. When this is tapped in upon it becomes another series of steps and evaluations. Then shall man truly say he has touched the spirit; with his mind, of course, with his heart, we hope, and with his scientific application.

In this respect we use the term *scientific* to mean controlled experimentation whereby the permission is given from the higher Selves: the soul evolvement of the one experimenting or issuing commands; and the one cooperating, which would be the instrument or individual who willingly could see the benefits to soul advancement.

RELATIONSHIPS

In this area one will become closer attuned to those of other vibrations and other areas of species development in relationship to man's concerns; and through that will see, and will have proof, finally that there is no crossing over from one species to another. For it is here, in the recording box of the brain matter itself, that the great differences are significant and observable in one species from another. It is here that the consciousness varies and the realizations and the levels of evolvement vary, because it is here within the brain that one can observe and dissect—not physically dissect, but spiritually manipulate and record through certain mechanical instruments, and thus dissect—that difference between man, animal, vegetable and mineral kingdoms.

There is much more that can be stated for the record, but it is not essential that you know it at the present level of discussion. It is essential that the information be recorded, within this particular series of revelations, that all things begin and end with the brain in the physical relationship to mind, spirit and soul essence.

12. DEVELOPMENT OF BODIES

SUN TEMPLE RECORDS

With this we conclude our phase of the embryological development of man in human form as evolving in a third dimensional frequency through to his original and infinite state of fourth dimensional life and being. This is Lord Uriel in conjunction with Nada, who is our channel via the instrumentation of her body and vehicle, Yolanda of the Sun Temple.

It is necessary to explain that phrase in order that you comprehend why all this information has been obtained in, and held in, etheric frequency form in the records of the Sun Temple, here upon the planet last evidenced in Atlantis. When Nada-Yolanda incarnated on the Earth in that embodiment as the high priestess of the Atlantean temple known as the Sun Temple, she had unfolded all this information for man in regards to his scientific developments and experimentations. Therefore, it is quite appropriate and necessary that she record them once more for the final days of America, or that civilization which shall be recognized as the final third dimensional frequency expression.

EXPANSION OF CONSCIOUSNESS

As mentioned in the prior chapter, all of light is incorporated in the brain cellular structure. It is here that the spirit controls the physical, the astral and the emotional bodies as well as the mental apparatus through which mankind can work upon the Earth planet. You also have been given that frequency modulations are separate and unique in each cell within the body. And within each of the bodies of mankind operating on the physical dimension, his similar or corresponding areas operate on a slightly different dimension or frequency modulation.

It is not yet common knowledge or acceptable development on the planet Earth how these frequencies can be recorded. But among your scientific expressions and experimentations are some who can record this in equipment called radionics. Hieronics is an instrument, using the electromagnetic frequency, that can record the degree of spiritual light and can help to maintain that spiritual or etheric light manifestation in the physical body.

Via the electromagnetic equipment that shall be introduced before the end of this century on the physical planet itself, you will be able to help raise or to change those varied areas that need alignment in each of the bodies that relate to physical manifestation upon the Earth planet.

It also has been given to you that any method other than the spiritual light force is not considered the highest form, or the most desired method, by which man can reach and sustain his light form on Earth. This still remains true. However, in the latter days we can help in sustaining this light frequency via those recommended and approved methods that cannot supersede any of your own spiritual development but can be aids to that development when under the control of masters or Hierarchal Board decree.

It is not so in the lesser systems that have been mentioned prior to this. For they force or they demand a certain reaction, or in some instances they do not work with the control or the approval of the higher Self or the light body involved. We wish also to make it quite clear that many attempts to manipulate brain expansion via breathing, meditation, drugs or any other such system can be considered quite harmful and not at all soul permanent.

For the soul aspect of man is that part which denies the overt or compulsory method of expansion; and would resist it, in most cases. In other words, your conscious application and acceptance of these things are on one level or one degree of your evolvement and development, but do not include necessarily the soul aspect, which is that part of you that is necessary extension and necessary counterpart to your proper evolutionary growth throughout all the systems in which you do expand and function.

NEED FOR SOUL ASPECT

Many have wondered how each of these aspects has been developed or has been incorporated into the proper relationship and proper evolution of man on Earth and throughout this solar system, without giving enough credence to the separate vibratory frequency rates of the various levels of consciousness. We shall attempt to outline very generally and simply the progress of this in the separate accounting.

Man became involved in the physical dimension when he had in himself only the light frequency or spiritual essence to sustain him in form and in matter. This was the period we mentioned as being as much as two hundred and six million years ago, if we could record time in the manner which you now do so upon the planet.

All these many factors involved in his engrossment with subtle form in his command and generalship were of such a nature that he could and did partake of them, in a manner of speaking. That is, he could absorb them into his system, or he could descend into them and participate with them, via the fact that all things are created out of divine or spiritual essence to begin with.

In some cases this lasted weeks or days or hours of time, as you may record or think of it at the present time. It was not essentially an incarnation, as mentioned prior to this, but an experimentation and an effort to raise or to participate in, or to exist with, that which is of a subjective nature and under one's own subjugation.

When this became a habit pattern within the race of man in this fourth dimensional frequency body, he then had to exist with it or to incorporate it into his entire memory pattern. It was at this time that the soul aspect or the emotional nature became a necessary factor, as far as his experience within the solar system was concerned. For he could not memorize or keep in record form within his own conscious application all these varied calculations, experiences and so forth.

As is also mentioned prior to this, he had to keep in memory form all things of each one within his frequency development who participated in a like measure. In other words, everything that was experienced by any one individualization of the sons of God anywhere within the same frequency modulation had to be recorded and memorized and utilized in each succeeding step or experimentation.

CREATION OF EVE OR SOUL

When this became a concerning problem to all those evolving in this present concept, it was necessary to devolve a part of the Self or spirit and to make it into a functioning re-form now known as or called soul body. This we also have explained.

Out of the rib of man came the soul, or the Eve self. This too has been explained in part; and is recorded in many scriptures, not only that known as Genesis of the Old Testament. But according to the memory of those who have had this revealed or those who have been raised into the Christlike consciousness, such as Moses the prophet who wrote this book or parable for the race of man at that time, we again have repeated the concept.

Part of the spiritual essence or part of the Adamic race was removed and was formed or was re-formed into that which we call the soul essence, or the counterpart within mankind himself which can function totally and exclusively as that part of himself and be his own record keeper. In some cases, many have called it the conscience of man.

In this record or computerlike aspect, many have confused it with other essentials and many have regarded it as a form other than one's self, to which one becomes wed and therefore is never separated from it. In this respect we expect to correct the misconceptions and to put all things again in proper, rightful place and origin.

RISING ABOVE SOUL

It is this aspect of man that became his conscience; when it should have been the spiritual Self or the I Am Self that was within, the essence out of which all things are created, that should and must always act as the conscience or the inner voice to guide man. Therefore, out of this conscience based on a lower aspect and a multiple series of experiences and experiments, we have had resolved the tremendous evolutionary pattern that proceeded to confuse and to cause conflict within man as an evolving being upon the Earth.

You will remember, of course, that this aspect or part of man came out of the so-called rib or heart section, the solar plexus section, of his development. It is because of this that often the heart leads the head; whereas the head, wherein is the brain, has the capacity to receive the light and to hold the light and to record the light, and should be the factor that controls his movements, his motions and his many decisions as regards his own evolution and conduct upon the planet Earth and throughout the solar system.

It also was symbolically recorded by the Master Jesus upon the cross at the time of his crucifixion, when the sword was plunged into his soul in order to release that soul and to hasten his ascension or resurrection expression. You have no need to be given a great many details regarding that as an aspect.

But if you will correspond your understanding with the fact that Adam, or the Adamic race or the Adamic function in the race, was the beginning of man participating in the physical dimension as a fourth dimensional being in a third dimensional plane, and Jesus the Christ exampled or represented the final step or stage of man's evolutionary progress as a fourth dimensional being expressing on a third dimensional plane and showing the way to return into total fourth dimensional frequency consciousness and expression and demonstration, then you can see why we are recording here the beginning of the soul, taken from the rib, to the point where the sword thrust into Jesus' side represents the release of that soul out of the third dimension, allowing the fourth dimensional or Christed man to rise above the soul and to become again the resurrected or light-body manifestation, which he demonstrated so perfectly after the crucifixion in the period called the resurrection, through the period which we call the ascension from out of this plane of operation altogether.

MAN'S SEVEN BODIES

As man then developed his spiritual Self through the brain expansion on the third dimensional level, and used it to hold the light and to record the light functions of his God Self, and became a part of the Earth through the release or development of a soul part of himself—or his mate or counterpart being his feminine nature or emotional nature—he then began to develop other functions or bodies also, which we have recorded on this plane and planet in order to aid him in evolving back to the one perfect body which can control, or which can release him from, all third dimensional aspects.

These other bodies have been recorded by other sciences of a spiritual nature or in other literature concerning this type of subject. They are the physical body, the emotional body, the astral body, the mental body and the etheric body, which is related to your light body and is your light body. Beyond this you have the magnetic force field, which alone is God force and concerns only your relationship with all essence and matter.

No one on this planet has been able to achieve his relationship in the seventh body. Therefore, we cannot give you the information on this plane regarding that, for the present time. For until Jesus, the master of this plane and planet, can return in his demonstration of redescent out of ascension, and lead the entire race consciously into its resurrected form or return of light-body function, and then ascend out of material matter and control material matter as he did in the resurrected period and in the ascended state in which he is at present, this information has to be withheld to a large degree from those who are achieving a school and a program by which you can raise all men and all matter into the original or final step of fourth dimensional frequency.

Each one of these bodies or aspects of man has operated, and will continue to operate, at a slightly different frequency level. These various levels will be recorded and will be shown as being influential upon the physical, mental and emotional bodies as well as the mental status. But where the mental status can contribute to reunion with the spiritual status or the light-body function is through the projections or use of mind power.

Too much does man dwell on his mental-intellectual aspect of mind rather than on his spiritual-mental, which is the mind power used by the spirit through the mental apparatus; which, again, is your brain concept and function. The brain can be used in part for great intellectual comprehension and expansion. But where the brain has been in a state of suspension from expanding to its further functions has been in the comprehension and acceptance of the spiritual part; which is connected with the spinal cord, and releases in that section of the pineal and pituitary glands—one representing the brain, the other representing the spine—to combine in a fully manifested spiritual-mental-physical evolutionary progress.

Incidentally, it is the pineal that represents your spiritual inflow and the light emission, and the pituitary which is the corresponding physical or mental aspect that receives that emission of light or submits to the light frequency as it records itself upon the physical frequency, or the rate of frequency, which your particular body represents.

INDIVIDUAL GROWTH

No two bodies are alike, although there is a comparable rate that expresses similarly in each body. Therefore, although the representative cells or organs of two or more or all bodies of man are similar in rate, there still is the very slightest degree of difference. This can be observed only via the spiritual consciousness of the individuals involved. This means it takes a master consciousness, one raised sufficiently into the light-body consciousness, to discern that smallest fraction of difference between one physical vibratory rate and another physical vibratory rate, where it concerns one individual versus another individual.

It is for this reason also that we have asked you to understand and to consider the fact that under no circumstances can anyone but a master consciousness really change, or help you change, the frequency of your body from physical or mental into spiritual consciousness, or to dissolve whatever error conditions exist in any of the lesser bodies which you are concerned with in the process of evolving from the finite into the infinite rate of exchange.

Therefore, if you still insist upon using minor means or temporary means you will not be progressing, but in some cases you will be thwarting your own growth; because, as we already have explained, for every mistake that is sustained and impressed upon the soul body, the individual has to go back over the exact same road or routine in order to come out and to find himself at the crossroads where he made the error condition or consideration.

Therefore, you do well to work only on one aspect of your evolution and growth on this plane and planet in the present time. That is to seek and to sustain spiritual light in and through your entire body, and in and through your mind and mental faculties, and in and through all soul evolvement and considerations, as it concerns your present and your past relationships in the growth process.

RACE GROWTH

You must take into consideration all who are in the same area of development, which means all the race of man evolving simultaneously. For you cannot have the growth pattern extended for one, and the others within that race be excluded from it.

This is further exampled by the Master and Prince of this plane, who is called Sananda but is known from his last incarnation as Jesus of Nazareth, in that he no longer is concerned with himself and his own evolution, but since and always has been concerned with that progress of man as a whole upon this plane and planet.

This material or explanation is contained in other literature and facts portrayed via this immediate channelship and others similar to this service, and cannot be expanded upon at the present time. For it is recorded sufficiently and is understood sufficiently on this dimension, as to the service and function and return of Sananda, who is Jesus the Christ of Earth. You will be given sufficient evidence of these concerns as the time progresses in the planetary program we call the latter days or the Mark Age period and program.

You have much to dissolve regarding your present embodiment and the race's series of embodiments, in regard to evolutionary progress on the planet for the last two hundred and six million years. Beginning with the sons of light or the Elder race encouraged to participate in and to supervise those life forms upon the Earth, until the present time when only a few are aware of the fact that their light bodies can influence the progress and be reinstituted upon the Earth planet, you have a wide difference of opinion and concepts.

Therefore, it shall take a great many years, and even an entire age of two thousand years more, to bring about the total restitution of the light bodies in manifestation on the third dimensional frequency. Meanwhile, the third dimensional frequency of Earth is changing, because as each person comes into and expands his light-body influence upon the Earth it slightly changes the frequency of those other forms upon the planet, in the other species' developments as well as in the race itself.

More predominantly, of course, it affects the race within which it occurs. Therefore, the race of man will be accelerated in its growth and evolutionary comprehension and in its expression as a physical entity upon the Earth planet. But as each one does expand and grow in this manner, he affects many more with whom he comes into contact consciously, subconsciously and superconsciously.

CONTROL OF INFLUENCES

Via this method, those in conjunction with one another or co-ordinated with one another can influence all the other forms or formations of matter—or species, as you call them—in the planet itself, thereby changing all frequency rates upon the Earth, from the mineral kingdom up through the vegetable, the animal and into the human expressions.

Everything you come into contact with actually influences you, in one form or another. The lower down the scale that you relate to, the less opportunity you have of raising yourself into the highest form you may desire, which is the spiritual or the light-body manifestation.

For concomittant with that light-body manifestation you have many faculties, talents and expressions which control the third dimensional frequency, and naturally would be very desirable; for it contains with it many rewards and satisfactions and controls over the environment, over the life and over the mental processes with which you always are concerned as a growing, experiencing and adventuresome type of species yourself.

Therefore, be very much concerned with those forms of life with which you concern yourself and with which you relate in your daily life and experience. Be discriminating and take upon yourself only those experiences and emotional contacts that can aid in this development and can help to change the frequency modulation of all your bodies simultaneously.

This means a great deal of mental, emotional and especially physical control as regards your body, your mind and your astral frequencies. In this respect you shall grow and shall know from within all that has been contained herein as firm truth and factual evidence that can be demonstrated. For here is the key to all truth: if it can be demonstrated and you can see the fruits of that by which you have applied yourself, then you can accept it as cosmic truth and divine law in operation.

However, if it does not demonstrate for good and does not evolve of itself to a higher step and a more conscious application of love and light and law in action, you have experienced additional error and further confusion from which you will have to evict yourself and the entire race and all life form upon the planet, at some time or another.

This is a grave and complex series of statements. Yet, man within himself has the comprehension and the desire to assist himself, his fellowmen and all other life forms. By knowing the truths that will set him free, he will apply them, in time. But if it will be in time for his own saving grace and that of all who are concerned with the planetary growth and process is a matter that he individually and as a group race consciousness will have to decide in the next few years.

MAJOR EVOLUTIONARY CYCLES

It is that serious and it is that imminent and of immediate concern for man individually and as a race consciousness. For in the evolutionary process and in the procedural grants given to man as a race, we have reached the end of several cycles.

We have reached the end of a two hundred and six million year cycle which involves graduation in and around the central sun from which we have originated. It involves a twenty-six million year cycle around which this planet has been evolved within this particular solar system and the sun with which we all are involved, or the Sonship with which we all are involved. It also completes a twenty-six thousand year cycle from which this planet had been given its last and final opportunity by the Hierarchal Board of this solar system to do so.

We will have one further concluding statement to make about the Hierarchy and how it operates and why it must function as it does, when we have concluded this statement.

Therefore, man has come to the ending of three major and essential cycles: two hundred and six million years, twenty-six million years, and twenty-six thousand years, each of them concerning the completion of an entire orbit or octave by which man could master his own destiny and come into his own proper relationship with the whole cosmic unfoldment.

It is not a coincidence that each of these sums equates an eight. For in the eight cycle of time we have been given every opportunity to complete above as well as to complete below; in other words, to complete spiritually and psychically that which we have completed physically, emotionally and mentally.

Therefore, if we have not completed all these cycles we must begin elsewhere to do so. The fact that three major cycles concerning this race upon this planet, within this solar system, and for this part of the galactic contribution are ending is no mere happenstance. It is of extreme importance to recognize that there is only one area of time or one segment of time whereby all three levels of development within the planet, the solar system and the galaxy could come to the same concluding or same final octave completion.

This is why it is said to be of immediate and imminent concern to the race of man, and especially those who are of the Elder race or those who are of the light-body consciousness already but are unable to maintain and to sustain the light body upon the physical dimension at this time. It no longer can be delayed. In more concise terms, a period of two hundred and six million years has been given to the race of man to sustain, and to function in, a light-body form upon a third dimensional vessel or Earth expression in this particular solar system.

UNDERSTANDING FOR ACTION

The fact that there have been trials and errors and repeated civilizations given to this experimentation has been outlined very briefly and concisely here. All the history and the detail of the history are not of extreme importance, for each person will remember that part which he played in his own existence and evolvement sufficiently to confirm what has been given. Therefore, the giving of all the details through any one channel is of a lesser method than the more developed method of asking and knowing and giving each one his own unfoldment within himself, his own realization within his own consciousness, his own cleansing within his own soul and spiritual development.

This is precisely why the information has been given in the manner in which it has—some of it rather subtle, some of it rather direct, some of it very pointed and some of it rather obtuse—so that each one in his own consciousness may find that part which pertains to his disgrace, or fall from graceful state of life, and thereby control and cleanse it and contribute to the entire race progress. For as each one does contribute his own cleansing and rebalancing within his own individualized consciousness, he does contribute to the race consciousness and to the evolvement of the race within the cycles of time herein mentioned.

It is most important that all these concepts be understood first, then acted upon secondly. If they are not acted upon by a contributing factor within the racial structure, we have not gained the proper momentum for the frequency change to be made manifest in a sufficiently strong and meaningful manner. This requires a hundred and forty-four thousand demonstrating-souls to sustain the light-body function and frequency within their spiritual life, their soul expansion and their physical transmutation.

It is a corresponding nucleus within the race that relates to the corresponding cells within the brain itself; a hundred and forty-four thousand separate parts within the brain that can control the entire physical, mental, emotional and spiritual expression of each body as it relates to the third dimensional form through which it is expressing in this time and place within eternal time and space.

You have been given a great many factual decisions upon which you can act or react. According to your degree of reaction you can come to some formula or conclusive form of manifested procedures to bring about the desired results in the scheme or time schedule that has been outlined here. That schedule involves the period of time leading up to and ending this particular century with which you now are involved and through which you now are evolving.

There are many souls who will be brought into conscious contact with this information so they can act upon it. When they do, the reaction of many who cannot seem to accept or to relate it in their own consciousness can be aiding or helpful in this regard by the time the cycle of time is completed; by the end of this century, that is.

USE OF INFORMATION

You will have much more information to glean and to study via this method if you will apply yourself and if you will give yourself the opportunity to do so. But as said before, to repeat all the information at one time that already has been manifested in physical form would be a wasting and a repeating of divine essence and divine projections. This you will find is never done, to any appreciable degree or length.

As far as man is concerned, he can repeat a great many facts in a great many forms and still not comprehend them. But as far as Spirit is concerned, It records Its facts and Its impressions upon any area of Its faculties or functionings or forms. And, without repeating, this exists eternally; for it cannot be changed or dissolved, regardless of what the lesser manifestations or the confusing, conflicting thought patterns or thought forms do with it.

So, in this sense man has sought the truth and has sought the facts of truth many, many times and has considered the error conditions more seriously than he has considered the truth. By overshadowing the truth and by confusing the truth factors he has made up or has fictionalized many things that could have aided him all through these various procedures and plans that have been evolving and have been developing throughout time and throughout recorded history; or memory recall via his myths and his legends, which predate anything that is recorded, and merely are used for symbolic purposes at the present time and in the present state of his evolutionary growth and development.

Until we return with a final and lasting statement of the nature of these series of informational discourses, we can leave the present body and form which is the channel through whom we record these facts, and can leave you who are developing within yourselves a concept and a cord of knowledge through which you can weave a new testament and force for life embodiment and thought control upon this planetary expression.

So be it. In divine truth and devotion we give all that is within our capacity and function. But beyond this step each man is his own responsible savior and messiah. All is one, and through one all can be assisted, raised and thrust free. I am Uriel, speaking through Nada, the Yolanda channel of this spiritual individual. Amen.

13. REEVOLUTION

I AM SELF

I Am that I Am, which is to say: Jehovah, the everlasting flame, the bush that can never burn, the Father to whom you address your physical consciousness. It is out of the essence of God, the Spirit of all things, that I Am as its eternal place and action. You may never be without the reach of this I Am Self or consciousness. You have been created, on every level of your expression, from the substance and matter of this eternal flame within, the light of God.

I Am that I Am is wherein you seek to begin and to end all evolutionary processes. It is from this essential beginning that all levels of development have been created for your experience and know-how and beingness, whether on the highest reaches of evolutionary scale or the lowest of the kingdoms in God's many mansions.

You now reach up from where you find yourselves. But only through the I Am can you have any strength and success. Begin at the beginning and learn how this is done, so the ladder you climb is strong and each step has meaning and purpose, and you will not falter from the heights you reach in order to come into your own fulfillment as a son of God and a light worker through all dimensions and existences. When you began as an idea and an individualization of the creation of Son light and form, you descended rung by rung on this same ladder. Therefore, in the reclimbing of it you will have to know what is in practice and what is in practicability for the stable reliftment.

CREATION OF MAN'S LOWER BODIES

As man descended lower and lower into the elements of the kingdom of matter, he had to have assistance in each area lower than the one where he had been in conscious control and in relationship to his I Am Self or beginning nature. Therefore, an embodiment and an appendage were created from out of this essential essence, or light and fire, of God within. So, the multiple bodies of man were created, starting with the etheric form in which he knew no sin and no fault with God the Father, or the aspect of God which is the creation of His dual nature, positive-negative polarity in double or balanced harmony.

Triune being of God is the Sonship in relationship to the Mother-Father aspect. The etheric form is that which gives it individualization and perfect seed to begin its experiments independently and with free will. As man climbed down the ladder, he had with him a calculator or a computer that was able to record all his independent thinking and findings. This is essentially the brain in man, which allows him to think and to know and to realize his unity with God the Creator and with all essence or form thereof.

In descending below this level, he had created an emotional body that was his soul. From this soul, which was taken from the heart or feeling nature of man, he had a recording device or memory bank that allowed all past episodes within himself and the race itself to be expressed and reset.

In the next step downward he had created an astral form. In this form he could descend into, and ascend from, matter in its nearest frequency to which he wished to have personal experience and exploration. This then became his entrapment. For the next step below was actually entering in, participating as a matter form in, whatever area was of such substance and materialization.

ONE WITH MATTER

Out of all these he has created for himself a cycle of evolution that is downward and leads upward. For the ladder of God or of Spirit is never one way. It must lead both down and up, and up as well as down. For man is part of God, and God is in all things that exist. There is nothing that exists that can be without the primary Source or fundamental Creative Principle in it.

Seek this common denominator in everything you touch, you see, and you feel from within. For there you will refine yourself and will become as one with it. But let not yourself become enmeshed in whatever you experience; for then you become that, and that is all you do find for yourself in the time allotted to you to fulfill or to fill yourself in it.

All this is given as a preliminary exercise for you to come to realize that man and matter are, in a sense, one and the same thing. Man the spirit is one with matter, for man experiences matter. Man the soul aspect is one with matter, for the soul has taken on matter in order to experience it and to sense it. Man the physical has put on the matter form and has become so enmeshed in it that he cannot see from beyond it that he is more than just the matter or the form.

But by recording these things we break the pattern and we set a new mold in man's thinking and consciousness so he has achievements beyond the matter range and will seek to exercise his divine right to portray himself in the true and highest sense that is within him; which is to say, the I Am that I Am.

Furthermore, it is given that man may come to comprehend and to accept the fact that all works in a divine and orderly fashion, with each step leading to each other step, and each creation having within it a relationship to all other creation. In addition, it is given so man may realize there is a hierarchy of evolution; in himself first of all, and in all combined relationships as a part of himself and his relationship with other matter and forms and beings.

SOCIETIES

Since no one is alone in the universe and since all things are related or are divided into one common barometer, which is the I Am and the spiritual essence of the I Am Self, then he knows that the things which shall be given according to these principles are divine and true and consciously applicable to his own life.

For no man can exist without his fellowmen. It is entirely impossible anywhere in the universe to be alone and without the conscious awareness of other entities or individualized consciousnesses participating in the here and the now, wherever you may be. No one may cut himself off from his fellowmen, regardless of what act he performs or what isolation he wishes to experience in any area of solid and sustained life frequency. This means that every area man has had to have developed, because of his issuance of a new area of experience and exploration, has within it a series, or a society, of souls or spirits who wish to contain themselves within that particular area or episode.

When man left his I Am Self to descend into personalized, etheric, individualized consciousness, so many were collected to do this. Many are still in that area or in the etheric vibration of our solar system and the areas beyond this solar system which sustain the individualized thought consciousnesses of sons of God or the light bodies of mankind as we have explained it, and as it has been evaluated, herein.

When it became necessary for him to have a mental and then an emotional and then an astral form, a soul body, he left within that area of consciousness those who wished to go no further and who thus have sustained societies and life and groups according to that level of experience. In other words, there are civilizations and races —to use it in a minor term—who are waiting for, and lifting themselves into, other and higher areas at all times.

REASCENSION PLAN

Those who have descended unto the lowest depths of society congregations, which we consider the Earth plane matter or third dimensional form, are waiting to be raised and are lifting themselves up one by one on the rung above each area, and experience the liftings-up or the assistance of those who have remained in, or who have grown up again into, that area.

As you have descended down, so you must ascend up. After you have ascended to the I Am consciousness you then return unto those areas, if it be your will, to aid those who have not awakened even unto that level of consciousness yet. For all are part of the One, and the one essence or light is in all equally. None can be safe or secure in his own position until all are awakened to the same essence and adventure of experience.

Therefore, it becomes necessary to form hierarchies, plans, governments, groups and brotherhoods unto a society or a scheme that will bring about a higher form of desire, or a higher goal, than exists in the present impasse, wherever it may exist. This is so unto all areas, without exception.

For there is none that has not a higher goal or a higher place or a higher plan than wherein it exists at that present time. Although all is in the present and in the now and in the awareness of the God consciousness within, so it is also true that there is a higher form that can be experienced; and a lower form that has not experienced what you are experiencing at the present time. This is the evolutionary pattern and this is purpose behind creation and the going out from creation to have knowledge of all other aspects of creation.

It does not follow that in every case those who have the I Am consciousness intact and have not lost the perspective of their I Am origin or source become entrapped in the area wherein they experience or desire exploration. But it does follow that wherein a soul or a series of souls does fall into such entrapments of the source for their experience and exploration, there is a plan existing for them to find wherewith they may remove themselves from that entrapment.

If it were not so, they would be without the source or the energy from within, which is the essence or the beginning of all life and life form. It cannot be other than this. For life is, energy exists. In this a whole series of developments occurs. But all that is is in matter. Therefore, in the matter form this isness, this source or energy must be urged, and must find a desire, to express itself and to shake off all its hallucination or matter materialized.

EXPLORATION

When man began to fall into the mesmerism of this matter form or life here on the planet Earth, a Hierarchy was formed; and had been formed even before this event. For it was the series or the society of Selves which could promote the various explorations. Certain assignments have been given, under the assumption of a Board of Directors, or as termed a Karmic Board by present terminology.

This Board is organized in order to send forth these expeditions of souls who volunteer under the freewill premise that they can and desire to know and to record and to bring about the information for all parts of the same species to conclude the remedies and the functions of their purpose for existence as sons of God, which is the governorship over all dominions and dimensions.

In this way whole congregations of souls or races—again using the term quite loosely—are put forth as armies would be put forth, or bands of workers would be put forth, to decide the best means for consciously exposing themselves to the dangers, and to the advantages, of that frequency form and development; of another species, perhaps, or of another area or of another dimension, or even of another galaxy that is just re-formed from out of Spirit essence and of projections of the balanced duality of God the Father and God the Mother.

So, it is up to the sons of God or the God Self, which is the trinity aspect or third aspect of God's activity that is manifested individualized identities, to know and to share and to experience it in its fullest extent. This is how hierarchies are born and how they are fed into a common computer in which all the sons of God can share and can know what it is to be in existence and to be part of the God Force.

This is the only species which has this privilege and knowledge. But all other species are part of the oneness or the God consciousness and the essence of Spirit, Which is all that can create and can be. All sons of God can enter into, and can participate with, every other form of creation in order to know it and to explore it and to be part of it in the sharing of this oneness and everlasting flame or light.

FEAR NOT BUT SEEK HIGHER

So, be not ashamed of that which you have experienced and be not fearful of that which is to come. For all is divinely organized and ordered, and not by some arbitrary board or by some arbitrary figure that can be wrought up or can be turned away from your pleas and cries for justice and mercy and love. All these things are the essence of God. They are in you eternally, for they are part of God and you are part of God. They are in all creation, for all creation is part of God. Therefore, this is the aspect which must manifest itself, if you will make yourself available and open to it.

Therefore, fear not, and be not worried, that your existence is doomed or limited; for it is not to be trusted, in that respect. All is forever in God's good hands, and can be made manifest in your experience and evolution if you will put your attention toward that one thing which is in all things and which is the higher aspect of the Creative Principle.

If you seek else, you will reap that for the time being. Unfortunately, we must face the fact that *the time being* very easily can mean a seeming eternity to you in consciousness, because you are suffering in one body or in one form or in one civilization or another. Thus it is to your best interest and intent to seek only that which is higher and better than that which you are experiencing and knowing in the present time. For without that goal you will be left in the lesser or the more difficult straits in which you now find yourself.

But should you put your attention on these higher matters, then those of a higher nature, those of that consciousness will be attracted to you and will seek you out as you are seeking them out. The two shall become as one. In other words, they shall aid one another and shall sustain one another and shall become as unified. For in all things there is unity in matter, there is unity in Spirit. In this you have your proper evaluation and destiny.

PARABLE OF ADAM AND EVE

When we, as souls individualized, sought out a common denominator upon the Earth, and that became the upliftment of all life form entrapped upon this planet, a great many schemes and plans were born. Among them was the one to enter into the materialized form and to lift it, through the descent into matter. This became the desire and the goal and the plan of the Elder race, who came from out of the central sun and descended rung by rung into the lower forms and energies surrounding this particular planet.

Those who had had experience of a dense vibration and who had lost their conscious awareness of the I Am identity within had been congregated together on this one planet to experience the lowest and densest form of vibration in order to evolve out of it. They were the ones the Elder race decided to raise and to return unto their proper evolution and gradual development of their higher energies and aspects, since they had been lost in matter for a great many eons of time in other places and other galaxies.

But from out of the central sun came this series of higher beings who knew that only by descent into the materialized form could they themselves begin to awaken from within the creative substance of this form. This began the race of Adam.

When it became necessary to keep the record so explicitly as to the rising and the gaining or the pros and the cons of the evolutionary experience, the Eve or soul for this particular area was provided. When the unification was complete and the aspect of higher consciousness or I Am Self was in comprehension and cooperation with the soul or record, feeling nature which we have called the woman in symbolic terms upon this planet, we created another series of episodes.

PARABLE OF ABEL AND CAIN

This was two factions in the one race or the one Elder brother relationship. One was known as the Cains and the other was known as the Abels, each working in separate matters and in separate areas to enforce this agreed-upon function. The descent into matter and the self-willed aspect which came about as a result of the descent into matter were the tasting of the knowledge of matter and spirit, or the turning away from the essential plan. So it, in its oneness, became two factions.

The story or the parable that has been created from out of this memory of the entire race growth is well taken and gives much to be incorporated into each one's final, future farings. As each one comes into the awareness of the one that is satisfied with the plan and the one that is dissatisfied with the plan, he can come to his own individual conclusions. Thus the war or the battle of Cain and Abel can be fought within himself.

But in the reality of it or in the history of it, as it recorded itself upon the planet Earth in that time so many eons ago, a series of individuals was committed to the race evolution along a set plan, and another was committed to destroying the plan or to subjecting the race to a further devolution of its experiences in the matter form. This has been the battle between the light workers and those of the dark or unenlightened consciousnesses.

The head of the Abels at that time was the soul known as Sananda. He, of his own vow and dedication, has asked to be the one to return unto Earth and to bring new and higher education step by step to those who are so enmeshed, including those who remain of the Cain group. So you have a series of evolutionary steps and a series of civilizations from era to era and age to age until the present time.

Among those who were on the Earth at the time was El Morya of the First Ray or the will aspect. He had with him his associate Kut Humi. These are etheric identities of individualized Selves or cells of God in the I Am consciousness and not necessarily physical personalities as you may understand or know them now.

But when the time comes for the light bodies to be remanifested on Earth, such as Sananda in the form of Christ Jesus—which was only one personality or projection of that single I Am Self or cell—you will see these light bodies, you will come to know these master consciousnesses who have dedicated eons of signs and wonders of themselves to this evolutionary process.

KARMIC BOARD

It is out of the few who could contain themselves and guard against the devolutionary scheme, who remain in the Hierarchal Board conclave or governmental setup in the present solar system; because, as you well can conceive, an entire solar system is involved whenever one part of it has become decrystallized or descends into deeper matter or deeper destiny.

The term *destiny* we can exchange with the word *karma*. This is the series of events which creates a reaction from a series of actions. When you set forth with a series of positive actions, you beget a series of positive reactions. When you set forth with a series of negative or false actions, or go against a set conceived and evolutionary plan, you beget a series of setbacks and failings which are reactions to that cause or karma you have set into motion.

So, the Hierarchal Board or the directors who set about to bring off these experiments had to create a Karmic Board. Among those who have been residing on that Karmic Board is the essence or soul spiritual identity of Nada, one through whom this spirit or I Am consciousness is permitted to have identity.

But it is not within the realm of personality, it is not within the realm of physical deeds and jurisdiction, it is not within the realm of soul memories or soul projections. For, as indicated and outlined here, each descent in the spiritual matter or spiritual consciousness begets a lower form of energy patterns or societies.

Therefore, the Nada aspect, for example, has experienced many soul episodes, many soul extensions. Among them is she who is known as Yolanda of the Sun. This is but a soul or individualized personality episode within the I Am consciousness or identity, the etheric form from which this race record is made up.

LEMURIA AND ATLANTIS

In this we have indicated that some souls had incarnated upon Lemuria, which was a civilization that completed itself approximately twenty-six thousand years ago, and some had formed a society or civilization known as Atlantis, which sprung up from out of the Lemurian civilization approximately two hundred and six thousand years ago. But in essence Atlantis failed also, from the primary premise or from the highest goal of its set plan. The final destruction of the land masses and the records of both civilizations occurred between ten and thirteen thousand years ago, as you now record time.

Thus your present civilizations, your present recorded records and memories are of the last ten to twelve or thirteen thousand years. Where all this was, in place or geographically locatable, has to do with the present series of discourses and the purpose for them in the present geographic area which is here called California.

As much as two hundred and six thousand years ago, when Lemuria seemed to have within its grasp the greatest potential for achieving the set goal of raising all third dimensional matter into fourth dimensional consciousness in a series of episodes and climactic developments, we were in most effective practice along the coastline of what is now thought to be, or related to be, your California land mass. We also incorporate many other physical areas with it, but use this approximate area as the key or the most civilized, the most developed potential area for the Lemurian civilization.

This included the area generally bounded by: Portland, Oregon; northeast to Glacier National Park in Montana; south through Idaho, the Great Salt Lake of Utah, the Grand Canyon and Flagstaff in Arizona; west through the Los Angeles basin; west two hundred and fifty miles into the Pacific Ocean from Los Angeles to Portland. This is the best that can be presented, as far as geographical reference points are concerned.

When the race became so entranced with its mental aspects, which was the predominant area of development of the race of man on the planet at that time, we had to abandon the project and the plan for the use of mental controls over the lower forms, including the nature kingdoms. The impressions implanted on the nature kingdoms, which means the devic and elementals, were such that more harm was being done than good. In other words, the balance became so overwrought in one direction of error that it was removed for the sake of cleansing the mental and preparing for even a lower descent, which then became the emotional or astral bodies.

Out of the kingdom then was a civilization that generally is called in today's expression the Atlantean period; which occurred on the east coast of this continent, from as far down as Honduras to approximately the New York area, and as far east as the British Isles and down through the Grecian isles, not in a full solid land mass but only scattered areas or land masses. The height of that civilization was the approximate area on your continent now called lower Florida, and out into the Caribbean Islands. This area was the seat of physical government as well as spiritual, educational government at that time. The height of that area or consciousness was twenty-five thousand years ago, and ended approximately twelve or thirteen thousand years ago.

Wherein this fact is solid and substantial matter or land-grant mass, we have a responsibility and an association of facts that cannot be ignored. The connecting land mass, or the final scheme or plan, then was to create a society or civilization that combined all areas of the Earth in one concerted ideology, governmental trial and test, as well as geographic land or physical relationships. This becomes the problem of the Americas. This becomes the challenge which is known as the Americas. This is incorporated in the governmental status and situation of one group called the U.S.A., or the New JerUSAlem.

I AM BROADCASTS

It is all consciously given, it is all consciously conceived, it is all consciously projected and it is all consciously unfolding. For it must be brought down unto the physical, that which has not succeeded on the lower mental planes and on the astral planes, and enforced now upon the physical planes.

Without much conclusive evidence and continued declarations, within the scheme and sociological structure I have given the benefit of the realization from the higher consciousness or the I Am Self projection into one individualized mind in order that it be broadcast out unto all who are of the like status and situation of I Am consciousness and achievement.

For what is given through the I Am awareness or identity of one incarnated physically and opened to the soul or emotional aspects, as demonstrated through the Yolanda aspect of the Nada Self, which is the I Am aspect of this individual creature, then is received and broadcast out unto every other one who is awakened and aware in the soul consciousness and who achieves, or strives for, an I Am identity and expression upon the physical plane and planet.

This is only the case in one individualized being. But when you have a congregation of such you have a society and governmental or sociological expression. For where two or more are gathered you gain the strength of broadcasting in amplification or multiplication.

ENDING AN EVOLUTIONARY EXPERIMENT

Furthermore, it is the granting of the awareness and memory of the last two civilizations; most unsuccessful, but in a sense quite successful in that they at least achieved predominant creative substance of another area or extension, or the I Am Self per se: the mental body and the emotional body in conjunction with the physical body so it could be joined up or linked up with the I Am consciousness.

These two factors—the Lemurian and the Atlantean, with the Lemurian expressing the mental body and the Atlantean expressing the emotional body—linked up with the physical or materialistic consciousness of the present American civilization make a very solid and congealed source by which all things can be unified and restated and reguided to project, to change and to enforce the I Am or Elder race conception in the approximate site where both became solid expressions and lower frequency developments.

It also was within the approximate site two hundred and six million years ago that the Elder race had produced much of its light-body energy forms and formulations for the growth and the expression of this particular planet. So, many thought forms also are in these areas that need restating and resetting for the higher evaluation and evolution of the entire planet and the plane of existence which is the third dimensional frequency rate of all life form and which is the final resting place of this experiment for the race of man within this solar system.

Those who cannot achieve this resetting and reevaluation during this lifetime or generation of lives, which means the next twenty to thirty years of life expansion on the Earth planet, will have to be reassigned elsewhere; but out of this solar system. For the solar system most emphatically and definitely has to move into a new place within the galaxy. The galaxy is but a part of an entire universe which is concerned with, or is an extension of, the energy coming out of the central sun.

All the souls and the individualizations who have participated in this race evolution and consciousness therefore must evolve with it and go on into new experiments and plans. Not one is to be lost; but not one is to be held back any longer from his evolutionary graduation status. That is the problem which now confronts the entire race of man, throughout history, throughout time and throughout all creation; of which I am but one part, and of which you through the I Am consciousness within you are another part and particle.

ONENESS WITH I AM OF ALL

Let us join as one within the One; which is the I Am Self or body within the Creative Force. For again here is a hierarchal and orderly governmental complex within the oneness of God; Who is but One, in the eye or center of Itself. But out of that we have the duality of God or the polarity of God, which is masculine and feminine expression, the Father idea and the Mother activity; with the Trinity coming from this duality or unification of the two aspects, which are really one, coming out and being individualized, or the Son light. From out of that Son millions upon billions of souls or cells are created with individualized identities and functions. No two are alike, no two have experienced alike.

Out of the millions upon billions of souls or Selves, which we call I Am identities within, are created many souls' experiences. Each I Am identity or individualization has had myriad types of expressions and soul episodes. Within those millions, many creative forces and many creative thought forms have been expressed throughout all the creations and dominions and frequency planes of existence in the many mansions which the one God, the one living Force, has created for us to have and to experience and to know and to recall and to command unto Him as our honored Father-Mother God.

In this respect we reach out and have memory within memory and we have relationship within relationship in ourselves and in each other. For as already stated, as one experiences, so all others may recall, or call upon the library of, that information and experience; or know or feel that which any other one has experienced. No two are ever alike and no two ever have experienced the same series of events or knowledgeable evidences. This is so throughout all eternity, from the beyond before and into the beyond afterward as well as in the present.

So, where you are you can know all things anyone else within the race of man has experienced. You can call upon your I Am presence or your spiritual essence and know all things that any other form of creation has experienced; not that you will live in that form as an animal or as a vegetable or as a part of the mineral kingdom, but that you can know all things, for you can read all things, being of the essence of God. The essence of God, being in all things, therefore responds to the essence within you, which is the One.

FOLLOW SPIRIT'S WILL

In this manner you can see how all things are evolved and developed and are made into an orderly confine of jurisdiction and law-fulfilling parts. But before you take upon yourself any one part and envy the part of another, you must be very, very careful to be sure it is the essence within you, or the I Am Self identity within you, which directs that desire and that mental activity.

For here is where the error of man has been and he has continued to fall into disrepute and disgrace; or to fall from perfect balance, which is the meaning of disgrace. He has desired his brother's or his fellow creature's experience or responsibility. Therefore, in the envy or in the gluttony of his unseeming desire or awe, he takes upon himself a series of actions and reactions that must be worked out through another series of reactions in order to get back to the first and proper, correct action or decision-making procedure.

Therefore, it is essential that in the reawakening of mankind now upon the physical embodiment—with the soul and the emotional aspects reawakened which were prevalent in the Atlantean development of his evolutionary cycle; and with the mental powers being reawakened, which was the episode of his Lemurian development upon this planet; and with the restating and reevaluating of his I Am Self or etheric body individualization being stimulated, which was the area and era of the light-body manifestation before these physical civilizations took place—we must begin to do all things in the right and proper design and scheme of Spirit's will, Spirit's desire and the essential overall good of God in man, God in form, God in the third dimensional frequency planet which is the Earth; evolving from that into a fourth dimensional frequency so this planet also may rise as a star in the system of the central sun. So be it.

In God's name, I Am that I Am. For only in God have I my existence, and am that. Holy is the name of the Father. The Father is the Jehovah consciousness or the messianic aura through which all physical embodiment becomes whole and reunited with the One, Which is Om, Om, Om.

THE SEVEN DIVINE ATTRIBUTES OR ELOHIM OR RAYS OF LIFE

ELOHIM OR IDEA	RAY OR COLOR	DIVINE ATTRIBUTE OR PRINCIPLE	ARCHANGEL IN CHARGE	CHOHAN OR DIRECTOR	MAN'S EARTHLY APPLICATION
First	Blue	Law or will; the word of God.	Michael	El Morya	Speaking the word for Spirit. Using sword (words) of truth for spiritual purposes.
Second	Yellow	Mind, intelligence, thoughts of God.	Jophiel (Hophiel)	Kut Humi	Intellectual understanding of divine laws, and wisdom in applying them.
Third	Pink	Feeling or love nature in God.	Chamuel (replaced Lucifer)	Lanto	Self-dedication and sacrifice to a cause, individual, country, religion.
Fourth	Crystal (colorless)	Development or manifested forms of God.	Gabriel	Serapis Bey	Completion or anchoring on physical of ideas and inspirations.
Fifth	Green	Integrating, unifying cohesiveness of God.	Raphael	Hilarion	Healing and synthesizing varied modes of expression. Unity in diversity.
Sixth	Violet	Transmutability or change of form through God.	Zadkiel	St. Germain	Cleansing, purification. bridging, old with new through transformation.
Seventh	White & Gold	Rest, peace, God's love for His work, love of His manifestations for Him. Love in action.	Uriel	Sananda	Completion of spiritual lessons on physical plane. Living in divine love, peace, rest.

MAJOR EVOLUTIONARY CYCLES

206,000,000 YEARS AGO

* Elder race of giants, fourth dimensional man in his etheric or light body, began experimenting with the elements and forms on Earth.
* Allegory, parables, myths and legends—such as Adam and Eve in the Garden of Eden, and all such scriptural references of most religions on Earth today, including such Greek and Roman myths —refer to this period of trial and error by fourth dimensional man with third dimensional substance and form.
* Etheric man helps to form land masses of Lemuria (Mu) by experimenting with the third dimensional elements of the planet.

26,000,000 YEARS AGO

* Final fall of man. Many of the Elder race became entrapped in third dimensional forms with which they were experimenting.
* Two schools of thought, the Cains and the Abels, developed. The Cains wished to dominate and to enslave the fallen, third dimensional subrace of man. The Abels believed they should teach the human subrace to rise back into the fourth dimensional form and to use their inherent etheric talents as sons of God.
* Caste system formulated: (a) those who maintained their fourth dimensional body and talents; (b) those who experienced third dimensional, physical sensations and remembered fourth dimensional or spiritual energies; (c) those who lost contact with their Sonship relationship and fourth dimensional form and were lost in physical or material embodiment and reimbodiment or reincarnation.
* Battle of the giants. Cains defeated the Abels, who were led by Sananda.
* Beginning of Hierarchal Board plans in our solar system to raise up the human subrace trapped on Earth between third dimensional form and astral planes.
* Lemurian societies or group civilizations began to develop, ap-

proximately from western U.S.A. through the Pacific Ocean and as far west as Mid-East and Far-East lands. Rise and fall of many groups, societies and civilizations.

206,000 YEARS AGO

- Hierarchal Board of the solar system, in the etheric dimensions, determined that controversy and continuation of the caste system started by Cains would not allow the Lemurian societies to achieve the goal of raising the entrapped race back to etheric forms while on third dimensional frequency.
- Elder race in fourth dimensional form withdrew from planet Earth.
- Contacts between third dimensional humans and fourth dimensional Elder race, only by etheric spacecraft and inner plane communion.
- Emotional and mental bodies of mankind in third dimensional form were anchored.
- Preparation of land mass of Atlantis, from east coast U.S.A. and Caribbean area to western Europe.
- Indian tribes, families, societies throughout South, Central and North Americas were the links between the Lemurian races and Atlantean groups.

26,000 YEARS AGO

- Height of Atlantean societies and civilizations. Attempt to equalize all.
- Peak of material, physical, third dimensional form for mankind on Earth.
- Determination by Hierarchal Board that superiority, domination, misuse of powers and knowledge would not allow Atlantis to succeed in spiritual goals.
- Beginning of breakup of land masses, both Atlantean and Lemurian areas; which sustained only remnants of its civilizations, such as the aboriginal societies of today.

10,000–13,000 YEARS AGO

- Final submerging of land masses and remnants; the period known as the Noahs and great flood.
- Beginning of recorded history of our present societies and civilizations.

- Atlantean cultures moved east to Europe and west again across the U.S.A., retracing steps of previous Lemurians two hundred and six thousand years ago who had moved east across the Americas to start Atlantis.

2000 YEARS AGO TO PRESENT: PISCEAN AGE

- Sananda, leader of the Abels, returned to Earth as Jesus of Nazareth to redemonstrate fourth dimensional or etheric talents in present third dimensional matter.
- Demonstration by Sananda as Jesus of resurrection and ascension as the pattern for all men to follow equally; no exceptions, no caste system.
- Promise by Sananda to return as Jesus from the ascended or fourth dimensional state in the end days or latter-day period now known as Mark Age period and program.

2000 YEARS AHEAD: AQUARIAN AGE

- Return of planet, mankind and all life form into the fourth dimensional frequency vibration; the etheric form, for the race of man.
- Interchange with man of other planets, planes and dimensions in our solar system. Brotherhood, unity, seventh step of spiritual achievement. Domination of Seventh Ray of Love, Peace, Rest.
- Spiritual government on Earth under the Prince of our planet, Sananda of the Hierarchal Board. Reentry into the federation of planets in our solar system, and exchange with the Hierarchal Board or Elder brothers.

GLOSSARY OF NAMES AND NEW AGE TERMS

Abel: son of Adam and Eve in biblical allegory; was not a person. Abel and Cain were the two clans of the Adamic race within the Elder race on Earth prior to Lemuria. The Abels wished to help raise their fallen brothers of the human subrace back into the fourth dimension. The Cains wanted to keep the humans in the third dimension as subjects.

Adam: first man on Earth, in biblical allegory; was not a person. A group of the Elder brothers known as the Adamic race. Sananda was one of the leaders. *See also* Adamic race.

Adamic race: those of the Elder race who descended onto Earth in combination of third and fourth dimensional bodies in attempt to raise their fallen brothers of the human subrace who had become entrapped in the third dimension.

akashic record: soul history of an individual, a race, a heavenly body.

angel: a being of celestial realms.

Aquarian Age: period of approximately two thousand years following the Piscean Age. Cycle during which the solar system moves through the area of cosmic space known as Aquarius.

archangel: head of a ray of life in this solar system. First: Michael. Second: Jophiel. Third: Chamuel (replaced Lucifer). Fourth: Gabriel. Fifth: Raphael. Sixth: Zadkiel. Seventh: Uriel.

Armageddon: the latter-day, cleansing, harvest, Mark Age period immediately prior to the second coming of Sananda as Christ Jesus. The era wherein man must eliminate the negativity in himself and the world.

ascended master: one who has reached the Christ level and who has translated his physical body into the light body or etheric body.

ascension: spiritual initiation and achievement wherein one translates the physical body into a higher dimension.

astral: pertaining to realms or planes between physical and etheric. Lower astral realms approximate Earth plane level of consciousness; higher astral realms approach etheric or Christ realms.

astral body: one of the seven bodies of man pertaining to Earth plane life. Appearance is similar to physical body. Upon transition called death it becomes the operative body for the consciousness, in the astral realms.

astral flight: a journey by the astral or soul body.

Atlantean writing: the mother of all languages on Earth. Given by the White Brotherhood from other physical planets, thus it resembles so-called space writings. Samples have been given through Yolanda and other channels.

Atlantis: civilization springing from Lemuria, dating from 206,000 to 10,000 years ago. Land area was from present eastern part of U.S.A. and the Caribbean to western part of Europe, but not all one land mass. Sinking of Atlantis was from 26,000 to 10,000 years ago; allegory of Noah and the flood.

at-onement: conscious unification.

aura: the force field around an object, especially a person. Contains information graphically revealed in color to those able to see with spiritual vision.

auric: pertaining to the aura.

automatic writing: a channeled communication by one from another realm written via control of the subconscious of the channel or instrument over the hands. May be handwritten or typewritten. Paintings or drawings can be done via such automatic process.

avatar: a spiritual leader for a certain period of time for all of mankind on Earth, or elsewhere.

bilocation: being in more than one place at the same time.

Cain: son of Adam and Eve, in biblical allegory; was not a person. *See also* Abel.

cause and effect, law of: as you sow, so shall you reap.

celestial: angelic.

chakra: a center of energy focus, generally located around one of the seven major endocrine glands, but which penetrates the other, more subtle, bodies.

channel: a person who is used to transmit communications, energies, thoughts, deeds by either Spirit or an agent of Spirit. Also called prophet, sensitive, recorder, medium, instrument.

channel, communications: one who is able to relay messages from this and higher planes or realms.

children of God: the race of man.

chohans: directors of the seven rays of life, under the archangels. First: El Morya. Second: Kut Humi. Third: Lanto. Fourth: Serapis Bey. Fifth: Hilarion. Sixth: St. Germain. Seventh: Sananda and Nada. As channeled through Yolanda numerous times.

Christ: a title indicating achievement of the spiritual consciousness of a son of God. Also refers to the entire race of man as and when operating in that level of consciousness.

Christ awareness: awareness of the Christ level within one's self and of the potential to achieve such.

Christ consciousness: achievement of some degree of understanding and use of spiritual powers and talents.

Christ Self: the superconscious, I Am, higher Self, oversoul level of consciousness.

conditioning: spiritual, mental and physical preparation of one's consciousness and bodies.

conscious mind: the mortal level of one's total consciousness, which is about one-tenth of such total consciousness. Usually refers to the rational, thinking aspect in man.

consciousness, mass: collective consciousness of race of man on Earth, all planes or realms pertaining to Earth.

contact: a connection with someone else on this or another plane or level of existence.

contactee: one of Earth who has had mental or physical contact and communication with those of other planes and planets.

coordination unit: designation and function of Mark-Age unit, Coordination Unit #7 for the Hierarchal Board, pertaining to coordination of light workers and light groups on the Earth plane for the hierarchal plan and program.

Creative Energy: a designation for God or Spirit or Creative Force.

crux ansata; or ankh: a cross with a loop as its upper vertical arm. Antedates Earth usage. Symbol used by heads of Saturnian Council for utilizing the power of Creative Energy.

death: transition from physical life or expression on Earth to another realm, such as physical incarnation on some other planet or expression on astral or etheric realms.

dematerialize: change of rate of frequency vibration so as to disappear from third dimensional range of Earth plane sensing.

devic: one of the kingdoms of God's creation of entities. Concerns the elemental or nature kingdom.

dimension: a plane or realm of manifestation. A range of frequency vibration expression, such as third dimensional physical on Earth.

disease: condition of dis-ease or disharmony.

Divine Mind: God or Spirit; in reality the only mind that exists, man having a consciousness within this one mind.

Djwhal Khul: Hierarchal Board name for one who has channeled through many on Earth. Incarnations here have included: John Mark, author of Gospel of Mark in New Testament; Wains, spiritual name for present Earth incarnation as James H. Speed, a former director of Mark-Age. Channeled through Gloria Lee as J. W. of Jupiter.

Earth: this planet. When referring to the planet, Mark-Age uses a capital E, since it is the only name for this planet that we have been given through interdimensional communications via Yolanda.

Elder race: those sons of God who did not become entrapped in the third dimension as the human subrace.

elect: one who has been chosen by Spirit and the Hierarchal Board to participate in the hierarchal plan and program, and who had elected so to be chosen. One of the symbolic 144,000 demonstrators and teachers for this spiritual program.

electric body: one of the seven bodies of man pertaining to Earth life. Known more commonly as the light body, the etheric body, the resurrected body, the ascended body. Resembles the physical body, but not necessarily of the same appearance. This body can be used by the Christ Self for full expression of Christ talents and powers.

electromagnetic: an energy combining electrical and magnetic forces.

electromagnetic beam: used by space visitors, via their equipment, to effect control over person or thing on Earth.

elemental: a being or entity of the devic or elemental or nature kingdom.

Elijah or Elias: the biblical prophet regarded by the Hebrews as heralding their coming messiah. Was an incarnation of Sananda.

El Morya Khan: Chohan of First Ray. Prince of Neptune. El denotes Spirit and the Elder race. Morya is a code scrambling of Om Ray. Khan is a Sanscrit term meaning king. No Earth incarnation since Atlantis (despite claims by others), until as Mark Age or Charles Boyd Gentzel (1922–1981), a director of Mark-Age.

Elohim: one or more of the seven Elohim in the Godhead, heading the seven rays of life; creators of manifestation for Spirit.

emotional body: one of the seven bodies of man pertaining to Earth life. Does not in any way resemble the physical body, but has the connotation of a vehicle for expression.

E S P: elementary spiritual powers, the definition coined by Mark-Age in 1966 to supersede the limited and nonspiritual usual meaning as extrasensory perception.

etheric: the Christ realms. Interpenetrates the entire solar system, including the physical and astral realms.

etheric body: one of the seven bodies of man pertaining to Earth life. Known more commonly as the light body, the electric body, the resurrected body, the ascended body. Resembles the physical body, but not necessarily of the same appearance. This body can be used by the Christ Self for full expression of Christ talents and powers.

Eve: wife of Adam in biblical allegory; was not a person. Symbolizes the soul of man, his feeling or emotional nature. Was created to keep the record of experiences in lower dimensions.

evil: in metaphysical or spiritual usage, means an error or a mistake.

eye, third: the spiritual sight or vision. Spiritual focus of light in center of forehead.

fall of man: sons of God becoming entrapped in the third or physical dimension of Earth 206,000,000 to 26,000,000 years ago.

Father-Mother God: indicates male-female or positive-negative principle and polarities of Spirit. Also, Father denotes action and ideation while Mother symbolizes receptive principles.

Father-Mother-Son: the Holy Trinity wherein Father is originator of idea for manifestation, Mother (Holy Spirit or Holy Ghost) brings forth the idea into manifestation, Son is the manifestation. Son also denotes the Christ or the race of mankind, universally.

federation of planets: coordination and cooperation of man on all planets of this solar system, except as yet man of physical and astral realms of Earth.

forces, negative: individuals, groups or forces not spiritually enlightened or oriented, but who think and act in antispiritual manners.

fourth dimension: in spiritual sense, the next phase of Earthman's evolution into Christ awareness and use of ESP, elementary spiritual powers. In physical sense, the next higher frequency vibration range into which Earth is being transmuted.

free will: man's divine heritage to make his own decisions. Pertains fully only to the Christ Self; and only in part and for a limited, although often lengthy, period to the mortal self or consciousness during the soul evolvement.

frequency vibration: a range of energy expressing as matter. Present Earth understanding and measurement, as in cycles per second, not applicable.

Garden of Eden: biblical allegory; was not one particular locale on Earth. That period when man functioned in both fourth and third dimensions on Earth.

Golden Age or Era: the coming New Age or Aquarian Age, taking effect with the return of Sananda around the end of the twentieth century. It will be the age of greatest spiritual enlightenment in Earth's history.

guide: higher plane teacher for one still on the Earth plane.

heaven: an attitude and atmosphere of man's expression, wherever he is. No such specific place, as believed by some religions; except to denote the etheric realms.

hell: an attitude and atmosphere of man's expression, wherever he is. No such specific place, as believed by some religions.

Hierarchal Board: the spiritual governing body of this solar system. Headquarters is on Saturn.

hierarchal plan and program: the 26,000 year program ending by the year 2000 A.D. wherein the Hierarchal Board has been lifting man of Earth into Christ awareness preparatory to the manifestation of spiritual government on Earth and the return of Earth to the federation of planets of this solar system.

Hierarchy, spiritual: the spiritual government of the solar system, from the Hierarchal Board down through the individual planetary departments.

hieronics: a higher energy not known to man of Earth from the time of Atlantis until its reintroduction via Mark-Age in late 1968.

human: those of the race of man who became entrapped in the third dimension on Earth, forming a subrace.

I Am: the Christ or high Self of each person. Jehovah, in the Old Testament. Atman or Brahman.

I Am that I Am: identification of the Christ Self with God.

incarnation: one lifetime of a soul; not always referring to an experience on Earth only.

Jehovah: biblical term for Christ or I Am Self.

Jesus of Nazareth: last Earth incarnation of Sananda. Christ Jesus, rather than Jesus Christ; for Christ is not a name but is a level of spiritual attainment which all mankind will reach and which many already have attained.

karma: that which befalls an individual because of prior thoughts and deeds, in this or former lifetimes. Can be good or bad, positive or negative.

karma, law of: otherwise known as law of cause and effect. What one sows, so shall he reap.

Karmic Board: that department of the spiritual Hierarchy of this solar system which reviews and passes on each individual's soul or akashic record. Assigns or permits incarnations, lessons, roles, missions for everyone in this solar system.

karmic debt: that which one owes payment for, due to action in this or prior lifetimes. Must be paid off at some time in a spiritually proper manner.

Katoomi: Hierarchal Board name for Lord or Archangel Michael. Titular head, with Lord Maitreya, of Hierarchal Board. Archangel at head of First Ray.

kingdoms: celestial, man, animal, vegetable, mineral, devic. Denotes a category of divine creation. Evolution is only within the same kingdom, never through the various kingdoms. Transmigration—incarnation of an entity in different kingdoms—is an invalid theory.

kundalini: spiritual force which rises up through the spine in the process of awakening the mortal personality to the spiritual consciousness and powers. Symbolically called the snake or serpent. The fire of the kundalini begins in the lower spine or sex center (chakra), rising gradually until reaching the head or crown chakra, where union takes place with the Christ Self.

Kut Humi: Chohan of Second Ray. Known as Master K.H. Sometimes spelled by others as Koothumi, but not his preferred spelling. Earth incarnations have included those as Aristotle, Lao-Tze, John the Beloved, Leonardo da Vinci. Known also as Babaji.

Lemuria: civilization dating from 26,000,000 to 10,000 years ago. Land area was from western U.S.A. out into Pacific Ocean. Final destruction was 10,000–13,000 years ago; allegory of Noah and the flood.

light: spiritual illumination; spiritual; etheric. Also, God as Light.

light body: fourth dimensional body of man; his etheric or Christ body; one of the seven bodies relating to Earth living; the resurrected or ascended body through which the Christ powers and talents can be demonstrated.

light center: a group of light workers functioning as a spiritual unit.

light worker: a spiritual worker in the hierarchal plan and program.

Lord: God; laws of God; spiritual title for office holder in Hierarchy; designation given to one who has mastered all laws of a specified realm.

Lord of the (this) World: designation of spiritual title for one who holds office of Christ for a certain area, such as Lord Maitreya for the solar system and Sananda for Earth.

Love God and Love One Another: the two laws which Christ Jesus gave unto man of Earth. The motto of the White Brotherhood, the light workers in this solar system.

Love In Action: the New Age teaching of action with high Self, action with love; the Mark-Age theme and motto.

MAIN: Mark-Age Inform-Nations, the media outlet for Mark-Age.

Maitreya: counterpart of Lord Michael. Holds office of Christ for this solar system. Master teacher of Sananda. Name indicates function: mat-ray, or pattern for Christ expression. With Michael, is titular head of Hierarchal Board; Michael as power, Maitreya as love. He is of the race of man in the etheric realm.

Mark Age: designation of the latter-day period, when there are appearing signs of the times to demonstrate the ending of the old age. Also, designation for the Earth plane aspect of the hierarchal plan. Also, the spiritual name for El Morya in his incarnation on Earth (1922–1981) as leader of the light workers here during the Mark Age period and program, as executive director of Mark-Age unit.

Mark-Age: with the hyphen, designates the unit directed by Hierarchal Board members Sananda, El Morya and Nada for the Mark Age period and program. One of many focal points on Earth for the Hierarchal Board. Coordination Unit #7 and initial focus for externalization of the Hierarchal Board on Earth in the latter days.

Mark-Age Inform-Nations: MAIN, the media outlet for Mark-Age.

Mary the Mother: mother of Sananda when he last incarnated on Earth, as Jesus of Nazareth. Twin soul of Sananda. Her Earth incarnations include those as Zolanda, a high priestess in Atlantis; and as King Solomon, son of David, mentioned in the Old Testament.

mass educational program: spiritual program to inform and to educate the world's population concerning the hierarchal plan and program of the Mark Age or latter-day period.

master: one who has mastered something. An ascended master is one who has achieved Christhood and has translated or has raised his or her physical body to the fourth dimension.

master ship #10: mother-ship spacecraft of city size which is Sananda's headquarters for the Mark Age period and program. Has been in etheric orbit around Earth since about 1885. Will be seen by those on Earth when time approaches for Sananda's return to Earth as Christ Jesus of Nazareth and as Sananda, Prince of Earth.

materialization: coupled with dematerialization. Mat and demat are a transmutation or translation from one frequency vibration to another, from one plane or realm to another. Translation of chemical, electronic and auric fields of an individual or object.

meditation: spiritual contemplation to receive illumination, or to experience at-onement with Spirit or one's own Christ Self or another agent of Spirit, or to pray or to decree or to visualize desired results.

medium: a term for a communications channel; a sensitive; an instrument; a prophet (the preferred term in higher realms).

mental body: one of the seven bodies of man pertaining to Earth living. Does not look like a physical body.

MetaCenter: the headquarters of Mark-Age unit. First two were in Miami, Florida, beginning in 1960. Coined from meta for metaphysics and for Lady Master Meta, guide of light groups on Earth for the past 2500 years. A center for the practical study and application of metaphysics. Interdimensional headquarters for Hierarchal Board externalization on Earth. Mark-Age, Inc. is the legal vehicle for Mark-Age, filed in December 1961 and legally recognized March 27, 1962.

metaphysics: spiritual meaning is the study of that which lies beyond the physical, of the basic spiritual laws of the universe, and the practical application thereof in daily life on Earth.

Miami, Florida: approximate site of spiritual capital and Temple of the Sun on Atlantis. One of the last spaceports on Atlantis to welcome man from other planets. Original location of Mark-Age unit, 1960–1980. Name spiritually significant: M-I-AM-I.

miracle: a spiritual manifestation, or a work. There are no so-called miracles possible, in the sense of circumventing a divine law.

mortal consciousness: the awareness of a soul during Earth incarnation, prior to Christ consciousness.

Moses: the biblical Moses of the Old Testament. An incarnation of Sananda.

mother of God: denotes demonstration of Mother or negative polarity aspect of God. Also symbolizes the soul or the subconscious aspect of man.

Nada: Co-Chohan, with Sananda, of Seventh Ray. Member of Karmic Board of Hierarchal Board. Present Earth incarnation is as Yolanda of the Sun, or Pauline Sharpe, primary channel and executive director of Mark-Age.

negative polarity: refers to the female principle in creation. The rest or passive nature, as complementing the positive or action polarity.

New Age: the incoming Golden Age or Aquarian Age. Actually began entry about 1960.

New JerUSAlem: the United States of America will become the spiritual pattern for implementing spiritual government on Earth in the coming Golden Age.

Om; or Aum: a designation for God. Means power.

one hundred and forty-four thousand: the elect, the demonstrators and teachers of Christ powers during the Mark Age period and program. The number is literal, in that at least that number must so demonstrate to achieve the spiritual goal of lifting man into the fourth dimension, and symbolic, in that it does not preclude any number of additional ones from being included.

phenomena: manifestations which man of Earth considers abnormal demonstrations but which are according to spiritual laws.

physical body: one of the seven bodies of man for living on Earth. Has been expressing in third dimension, but will be well into the fourth dimension by end of twentieth century. The vehicle for mortal expression of the soul on Earth. The physical on other planets of our solar system expresses as high as the eighth dimension.

pineal: crown chakra or highest spiritual center in the body. When the pineal gland is opened and the spark between it and the pituitary (third-eye center) is ignited, Christ consciousness occurs.

Piscean Age: the period of approximately two thousand years now drawing to a close for Earth, to be followed by Aquarian Age. One of the zodiac signs designating a section of space through which the solar system travels around a central sun.

pituitary: the master gland for the physical body, spiritually known as the third eye. The kundalini fire must be raised to the pituitary center before the pineal can be united for awakening the soul to Christ consciousness.

plane: a realm, a dimension, a level of expression.

positive polarity: the male or action focus, as complementing the negative or female or passive polarity.

prince: a spiritual office and title, such as Sananda being Prince of Love and Peace as Chohan of Seventh Ray, and Prince of Earth as spiritual ruler of this planet.

Project Power: the noon prayer meetings held by Mark-Age since 1960, lasting usually only several minutes, for purpose of projecting spiritual aid to persons, places and situations.

prophet: in addition to usual meaning it is the term preferred by those of higher planes in referring to a communications channel.

psychic: refers to the powers of man focused through the solar plexus chakra or center. Not as high as the Christ powers.

purgatory: no such special place, but indicates the frame of mind or experiences of one not in the spiritual light and activity.

realm: plane, dimension, a level of expression.

reincarnation: taking on another incarnation, on any plane or planet, during one's eternal life.

Sananda: Chohan of Seventh Ray. Prince or spiritual ruler of Earth. One of Council of Seven, highest ruling body of the solar system. Previous Earth incarnations: Christ Jesus of Nazareth, his last one; biblical Melchizedek, Elijah and Moses; Gautama Buddha; Socrates, Greek philosopher; leader of Abels, in allegorical story of Cain and Abel; leader of Noahs, in allegorical story of Noah and the ark. Presently located in etheric realm, from whence he directs entire operation for upliftment of man and his own second coming; headquarters is ship #10, in etheric orbit around Earth since about 1885.

Saturnian Council: Council of Seven, highest ruling body of the solar system. Headquarters is on planet Saturn.

second coming: refers to each coming into awareness of his or her own Christ Self, and the return of Sananda as Jesus of Nazareth to institute spiritual government on Earth by 2000 A.D.

Self, high: Christ Self, I Am presence, superconscious, oversoul,

Atman, Jehovah. The spiritual Self of each individual. Differentiated, in writing, from mortal self by use of capital S in Self.

self, mortal: the spiritually unawakened consciousness of Earthman.

sensitive: a channel, prophet, instrument, medium. One who is sensitive to or aware of spiritual realms and occupants therein.

seven, unit #: number of Mark-Age unit, as Coordination Unit #7 for the Hierarchal Board on Earth during the Mark Age period and program.

seven in a circle: symbol of Sananda and of Mark-Age unit. Signifies completeness, wholeness and the step before spiritual manifestation. Indicates the seven steps of creation, the seven rays of life, the seven major spiritual initiations.

seven rays of life: the seven major groupings of aspects of God; the seven flames. First: will and power (blue). Second: intelligence and wisdom (yellow). Third: personal love and feeling (pink). Fourth: crystallization (colorless, crystal-clear). Fifth: unity, integration, healing, balance (green). Sixth: transmutation, cleansing, purification (violet). Seventh: divine love, peace, rest (gold and white). As channeled numerous times by Yolanda.

sin: error, mistake.

sin, original: man's mistaken belief that he can have an existence away from or be separated from Spirit.

sleep, teaching in: all persons are given spiritual instruction during their sleep state, especially in this latter-day or Mark Age period. Many are taken on trips, in the astral body, to other planes or planets as part of such instruction. Many perform works or do teaching while in the sleep state, via the Christ Self consciousness.

Son, only begotten: refers to the entire Christ body, which includes all of mankind, and not just a single individual.

Son of God: with capital S for Son, denotes the Christ body of all mankind, collectively. With small s for son, denotes an individual. All men are sons of God and eventually will come into that awareness, heritage, power and co-creativity with God.

soul: the accumulation of an individual's experiences in his or her eternal living. A covering or a coat of protection, over which the individual spirit can and does rely for its manifestations.

soul, twin: as an individual soul develops, it expresses in male and female embodiments. Eventually it will begin to gravitate toward either male or female for its expression in the Christ realms. While so developing, Spirit guides another soul toward the opposite polarity along the same path. Thus when one enters the

Christ realm as a male polarity, there will be one of female polarity to complement and to supplement, with the same general background and abilities. Thus each person has a twin soul. But this term does not mean one soul was split, to gain experiences, and then eventually merges back into oneness. Twin souls are two separate individualities at all times.

soul mate: one with whom an individual has had close and favorable association in one or more lifetimes. Each person thus has had many soul mates, but does not incarnate with or come into contact with all of them during any one lifetime.

Source: a term for God, sometimes called Divine Source.

spacecraft: flying saucer, UFO, unidentified flying object; common terms on Earth for interplanetary and interdimensional spacecraft of man from other planes and planets.

Spirit: God, Creative Energy, Creative Force, Divine Mind, Father-Mother God, Original Source.

spirit: the spiritual consciousness or Self of man.

spiritism: communication with those who have departed from the Earth plane. Term preferred by those in higher planes in place of spiritualism.

spiritual: term preferred over *religious* when referring to spiritual matters, as there are specific dogma and connotation attached to *religious*.

subconscious: one of the three phases of mind. Denotes the soul or record-keeper phase, which also performs the automatic and maintenance functions of the physical body. The relay phase between the superconscious and conscious aspects of one's total consciousness.

superconscious: the highest of the three aspects of individual consciousness, consisting also of conscious and subconscious aspects. The Christ, I Am, real, high Self. The real individual, which projects into embodiment via having created a physical body for such incarnation.

sword of truth: denotes the use of God's word and law to eliminate error, and to guide and to protect spiritual persons.

teacher, spiritual: one who teaches spiritual matters. May be on this or a higher plane.

teleportation: spiritual power enabling one to move from one location to another via dematerialization and materialization, without physical means. A Christ power. Symbol for this in Atlantis was dodo bird.

tests, spiritual: tests of one's spiritual progress and lessons learned, given by Spirit, by one's own Christ Self or by other spiritual teachers. Not temptings, which never are given anyone by any of the above guides.

third dimension: the frequency vibrational level in which Earth and all on it have been expressing physically for eons. Being transmuted into the fourth dimension, which was begun gradually by the mid-twentieth century for completion in the twenty-first century, but well into the process by the end of the twentieth. Does not refer to the three dimensions of length, width, and height, but to a range of vibration.

thought form: an actual form beyond the third dimension, created by man's thoughts. Has substance in another plane and can take on limited powers and activities, based on the power man has instilled in it through his thoughts and beliefs.

thought temperature: the attitudes of an individual or group concerning a certain topic.

trance: unconscious or semiconscious state wherein one allows the mortal consciousness, or at least the conscious aspect, to become inoperative. Many communications channels enter such in order to permit those on higher planes to speak or otherwise to manifest through the channel's physical body and abilities. A valid form of channelship, but not as highly evolved as the conscious channel, who does not give up conscious awareness during channeling.

transfiguration: a change of one's features, or of entire body, caused by overshadowing by one's Christ Self or by an ascended master.

transition: term denoting death of an individual on one plane so as to begin a new life on another plane. Also, general meaning of making a change.

transmutation: spiritually, refers to purifying one's mortal consciousness and body so as to permit raising into fourth dimension, physically and as concerns Christ consciousness.

trials: spiritual tests given one in evolution to see if lessons are learned or if obstacles can be overcome, as in training for a soul mission or any part thereof.

Trinity, Holy: Father-Mother-Son, Father-Holy Spirit-Son, Father-Holy Ghost-Son. The three aspects of God.

twenty-six million year cycle: a period of evolution for man in this solar system. The cycle since the final fall of man on Earth, during which the Elder race has been attempting to raise the human race that had become entrapped in the third dimension. Cycle to end 2000 A.D.

twenty-six thousand year cycle: the period of time, since the beginning of the fall of Atlantis, in which man of Earth has been given the last opportunity in this solar system for reevolution into the fourth dimension. The duration of a hierarchal plan and program to raise man from the third dimension into his true status as sons of God. Cycle to end 2000 A.D.

two hundred and six million year cycle: an evolutionary cycle for man involving graduation in and around the central sun from which we originated. The period during which man has experimented with life form on Earth in the third dimension. Cycle to end 2000 A.D.

two hundred and six thousand year cycle: withdrawal of Elder race from on Earth, and decline of Lemuria. Cycle to end 2000 A.D.

unit: a group of two or more performing hierarchal plan works.
Uriel: Archangel heading the Seventh Ray.

vehicle: denotes the body for one's expression, such as physical body.
veil, seventh: final veil separating man from knowing his divine heritage and powers.
vessel: denotes a vehicle or body for expression, as the physical body.
vibrations: the frequency range in which something is expressing; not in terms of cycles per second, or any present Earth understanding and terminology. Also, the radiations emitted by an individual, able to be received consciously by one spiritually sensitive to such emanations.

Wains: spiritual name for Djwhal Khul during his present Earth incarnation as James Hughes Speed. Name means WAy IN Spirit, or WAy IN Space. Demonstrating both of these aspects for man during the Mark Age period and program.
White Brotherhood: the group of spiritual teachers, guides and masters who, having discovered God's truth and having learned to demonstrate spiritual powers, are guiding those who have not into that same level of Christ understanding. Pertains to this solar system. Is not a formal or organized group. Also called the White Lodge, Ascended Masters Council.
within: denotes going into one's own consciousness.
world, end of: denotes ending of third dimensional expression for Earth and all on it, physically, and entry into a higher level of frequency vibration, the fourth dimension. The end of the materially minded world of man so as to begin spiritual understand-

ing and evolvement. Does not mean end of the Earth, but only entering a higher dimension.

wrath of God: the reaction upon man of his negative or error action contrary to divine law. Does not mean personal anger or action of God against man.

Yolanda of the Sun: present Earth incarnation of Nada as Pauline Sharpe, executive director of Mark-Age. Was her name at height of her Atlantean development, when a high priestess of the Sun Temple, located near what is now Miami, Florida. Also known as Yolanda of the Temple of Love on the etheric realm of Venus.

Zolanda: an incarnation of Mary, twin soul of Sananda, when she was a high priestess of the Earth Temple in Atlantis, on what is now central Long Island, New York.

MARK-AGE

Mark-Age has been commissioned by the Hierarchal Board of our solar system to be Coordination Unit #7 on Earth.

Therefore, we are interested in hearing of your spiritual awakening, development, experiences, goals, and desires to participate in the Mark Age period and program. (Mark Age with no hyphen refers to the period of time from 1960–2000 AD, not the organization.) This applies equally to individuals, groups, churches and associations dedicated to raising all facets of human life on our planet into expression of cosmic law and I Am Self consciousness.

Mark-Age United Staff donate their full-time services. Major source of income is love donations (US tax-deductible).

Love in Action is the Mark-Age motto. *Love God and Love One Another* is the cosmic law for the Aquarian Age.

We would be pleased to send you our free Introductory Catalog, *You Can Enlighten Planet Earth.*

MARK-AGE, Inc.
P.O. Box 290368
Ft. Lauderdale, FL 33329, USA